CHECKLISTS FOR MANAGEMENT, ENGINEERING, MANUFACTURING, AND PRODUCT ASSURANCE

VOLUME II

Electrical and Electronic Checklists
(Including Microelectronics)

Electrical and Electronic Checklists

(Including Microelectronics)

Volume II

In the series of manuals entitled

Checklists For Management, Engineering, Manufacturing, and Product Assurance

by

W. B. Rossnagel

SPARTAN BOOKS

New York : Washington

Library of Congress Catalog Card Number 78-133105
International Standard Book Number 0-87671-558-7
Printed in the United States of America.
Sole distributor in Great Britain, the British Commonwealth, and the Continent of Europe:

The Macmillan Press Ltd.
4 Little Essex Street
London WC2R 3LF

1	Checklists for management, engineering, manufactur	Used: Good	Worldwide from Puerto Rico, USA.Listing Includes Books Image . Please email me if you need to see more pictures! The orders are processed promptly, carefully packaged and shipped within 1 day of purchase. PLEASE NOTE! if you need the book quickly, please Purchase Priority Shipping. Media will not show updates in mail confirmation till reaches continental U.S.	$120.39

Packing Slip

Norman - Miller
HC 03 box 12365
Corozal, Pr, 00783

787-245-6650

onceuponatimebookspr@gmail.com

Ship To: ABE TP Orders
800 Avondale Ave
Columbus, OH 43212

[illegible]	[illegible]	[illegible]	[illegible]	Price

To my Mom and Dad—
for so many reasons

TABLE OF CONTENTS

RELATED CHECKLISTS

VOLUMES IN THIS SERIES

VOLUME I
MANAGEMENT CHECKLISTS—twenty-two checklists covering Computers, Configuration Management, Engineering, Financial, Labor Relations, Manufacturing, Marketing, Personnel, Product Assurance, Proposals, Publications, Small Business, Top Management, and other subjects.

VOLUME II
ELECTRICAL and ELECTRONIC CHECKLISTS—twenty-two detailed checklists on subjects including Cables and Connectors, Communication Radios, Multicoders, GSE/AGE/TSE, EMI/RFI, Lasers, Radar, Systems, Transmitters and Transponders, as well as other subjects. Also included are Microelectronics and Integrated Circuit checklists.

VOLUME III
MECHANICAL and ELECTROMECHANICAL CHECKLISTS—twenty-two checklists concerning Thermal, Stress, Actuators, Gyros, Servomechanisms, Relays, Tape Recorders, Solenoids, and other subjects.

VOLUME IV
PNEUMATIC and HYDRAULIC CHECKLISTS—twenty-two checklists on Pressure Vessels, Regulator Valves, Solenoid Valves, Burst Discs, O-rings, Hydraulic Motors, Vacuum and Hermetic Sealing/Leakage Rate, and other subjects.

VOLUME V
PRODUCT ASSURANCE CHECKLISTS—twenty-two detailed checklists on Quality (Inspection and Test), Reliability, Maintainability, Safety and Human Factors.

VOLUME VI
PROCEDURAL CHECKLISTS—twenty-two checklists covering a wide range of Production Control, Manufacturing, Purchasing, Design, Drafting, Cost Estimating, and other subjects.

VOLUME VII
SOFTWARE COMPLIANCE CHECKLISTS—twenty-two checklists on the myriad requirements for written justification of work accomplished, covering Configuration Control, Reports, Specifications, Planning, Drafting, etc.

VOLUME VIII
MATERIALS and PROCESSES CHECKLISTS—twenty-two checklists covering a wide range of materials and many significant manufacturing processes. These are typical of critical procedures and processes which can be effectively controlled by the use of checklists.

PREFACE

THIS MANUAL is Volume II in a compilation of checklists which the author began accumulating and using as a development engineer. Later, from experience in engineering-quality-reliability-test management and as an independent consultant, the author expanded the scope and detail of these checklists.

Included are selected checklists from other key engineers and managers. Combined, they form both a treatise on the use and application of checklists and a comprehensive and extremely useful reference manual for a wide range of products made by the aerospace, military, and commercial industries.

ACKNOWLEDGEMENTS

In addition to the contributing authors, acknowledgment is also given for assistance from the following persons in the preparation of this Volume II:

G. Adams	Motorola Semi-Conductor
R. Barnes	Vostron
J. Barney	General Electric
J. Campbell	General Electric
R. Cheetham	R.C.A.
C. E. Coon	Electronic Industries Assoc.
H. A. Coville	General Electric
J. D'Aquilla	Hamilton Watch Co.
R. Fellows	Hewlett Packard Assoc.
R. Fluharity	General Electric
W. Garrett	TRW
B. Hamilton	General Electric
S. D. Hiltbiddle	General Electric
M. Kelca	General Electric
R. Maslanski	General Electric
G. Noebel	Texas Instruments
A. Octaviano	General Electric
H. Powell	Powell Electronics
C. F. Schreiber	General Electric
E. Slaughter	Kearfott Division of Singer
W. Smith	General Electric
J. Stanton	General Electric
E. Tighe	General Electric
S. Webb	NARCO
C. Zierdt	Bell Laboratories
S. Zwerling	General Electric

And to the secretaries: Mary Lou Cini, Helen Sheridan, Doris Sidlow, Cecilia Villa, and Virginia Rossnagel

INTRODUCTION

THIS MANUAL, Volume II of *Checklists for Management, Engineering, Manufacturing, and Product Assurance,* is a compilation of twenty Electrical/Electronic Checklists, and two Microelectronic Checklists.

Each checklist consists of factors of utmost importance to program, project, design, development, production, reliability, and quality engineers; to designers, checkers, layout, and detail draftsmen; and to procurement, marketing, customers, etc., who require specific knowledge of systems, components, and parts.

A checklist can never be an all encompassing document, but should be used to assure that important considerations from past experience have been duly reflected in a new or modified design. A checklist is a form of "memory feedback system" which recent events have proven so necessary for the efficient development of reliable hardware.

On complex tasks, it is recommended the items in these checklists be checked off by the designer and/or project engineer and/or design review team as they are accomplished or verified.

The Table of Contents on the following page references a wide range of Electrical/Electronic components and systems, such as digital circuitry, communication radios, printed circuit boards, cable and harnesses, connectors, etc. It includes a practical and up-to-date section on integrated circuits and microelectronics.

The series of manuals entitled *Checklists for Management, Engineering, Manufacturing, and Product Assurance* provides a nucleus of checklists for the aerospace, military, and commercial industry, but in so doing is in itself a timely manual on the preparation and effective use of checklists as a tool in industry, government, and educational fields.

—W. B. ROSSNAGEL

Chapter 1

THE CHECKLIST CONCEPT

First, let us define the word *checklist.* The dictionary defines *check* as a "supervision of accuracy, efficiency," and *list* as a "series of names, numbers, etc., set forth in order."

Checklists per se came into vogue in the thirties for use by pilots in the DC-3 days of commercial aviation. Today, they form the basis of all training, standardization, and supervision of military and commercial flight crews.

In industry, aside from some special applications, the concept of applying checklists to special operations, tests, or inspections has primarily been a post World War II "missile age" innovation.

Commercial and military aviation, however, has spearheaded the use of checklists. Perhaps the prime reason, besides safety considerations, for their acceptance has been the necessity for standardized operating procedures among pilots, copilots, and flight engineers. With frequent crew changes, equipment substitution, etc., the need for checklists developed steadily.

Today, the airlines provide checklists for individual crew members and others to be kept aboard the aircraft. A typical modern checklist for both pilots and flight engineers follows. It is a TWA "Before Starting Engines" and "After Starting" checklist for Boeing 727-31 turbojet transport. It bears the F.A.A. approval date of April 29, 1969. Immediately following is a checklist from the DC-3 Flight Inspection Normal Operating Check List which has been in effect for more than thirty years on this pioneer twin engine commercial aircraft.

The fact that even seasoned senior pilots are required by the F.A.A., ALPA, and the airlines to utilize such standard operating

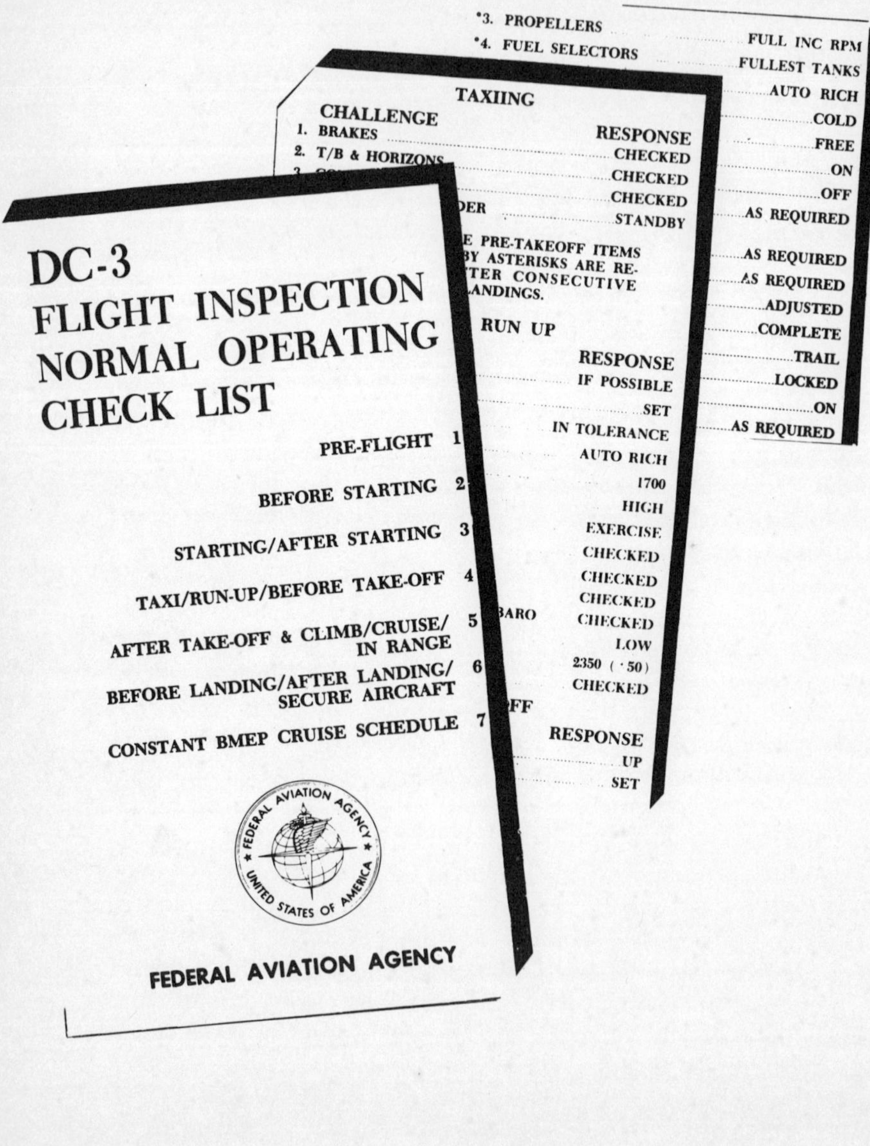
*3. PROPELLERS FULL INC RPM
*4. FUEL SELECTORS FULLEST TANKS
AUTO RICH
COLD
FREE
ON
OFF
AS REQUIRED
AS REQUIRED
AS REQUIRED
ADJUSTED
COMPLETE
TRAIL
LOCKED
ON
AS REQUIRED
TAXIING
CHALLENGE
RESPONSE
1. BRAKES CHECKED
2. T/B & HORIZONS CHECKED
CHECKED
STANDBY
RUN UP
RESPONSE
IF POSSIBLE
SET
IN TOLERANCE
AUTO RICH
1700
HIGH
EXERCISE
CHECKED
CHECKED
CHECKED
CHECKED
LOW
2350 (·50)
CHECKED
RESPONSE
UP
SET
DC-3
FLIGHT INSPECTION
NORMAL OPERATING
CHECK LIST
PRE-FLIGHT 1
BEFORE STARTING 2
STARTING/AFTER STARTING 3
TAXI/RUN-UP/BEFORE TAKE-OFF 4
AFTER TAKE-OFF & CLIMB/CRUISE/ IN RANGE 5
BEFORE LANDING/AFTER LANDING/ SECURE AIRCRAFT 6
CONSTANT BMEP CRUISE SCHEDULE 7
FEDERAL AVIATION AGENCY ★ UNITED STATES OF AMERICA ★
FEDERAL AVIATION AGENCY

Figure 1.

checklists is significant. Too often other skills and professions have looked down on the checklist as something not really needed. Some professions, if forced to write them, tend not to use them, and the checklists rapidly become obsolete, ineffective control documents.

Normal 727-31 Checklist

Before Starting Engines

At through stations (no mechanical delay or crew change), necessary to read only boxed items.

Pilots

1.	Gear lever and lts.	Down and Ck
2.	Parking Brake	On
3.	Flt. Instruments	Ck
4.	Altimeters and Clocks	Set and Cross Ck
5.	Brake Pressures	Ck
6.	Radar	Standby
7.	Transponder East	#2 Standby
	West	#1 Standby
8.	Start Levers	Off
9.	Exterior Lts.	Ck
10.	Emer. Exit Lts.	Armed
11.	Seat Belt, No Smoke	On
12.	Window Heat	Low
13.	Compasses	Slave
14.	Radios	On and Ck
15.	Flight Recorder	Ck and Normal
16.	Alternate Flap Sws.	Off
17.	Stall Warning	Ck
18.	Flt. Cont. Shutoffs	On
19.	Emergency Power	Ck and Off
20.	Pitot Heat	Ck and Off
21.	Start and Ign. Sws.	Off
22.	Eng. Anti-Ice	Ck and Off
23.	Wing Anti-Ice	Off
24.	Nose Anti-Skid	On
25.	Main Anti-Skid	Ck and Off
26.	Fire Control Panel	Ck and Set
27.	Oxygen Mask and Reg.	Ck and Set
28.	Static Sels	Both Norm
29.	Emer. Brakes	Off
30.	Comparator	Ck
31.	Horizon Sel.	#1
32.	Deviation Sel.	#1

Flight Engineer

1.	Battery	On
2.	Fuel Quantity	
3.	#1 Boost Pumps	One On
4.	Crossfeed valves	Open
5.	Hyd Quantity	Ck
6.	System	
7.	System B Pumps	Ck and One On
8.	Oil Quantity	Ck
9.	Bleed Valves	Open
10.	APU Panel	Ck
11.	Gear Pins	removed

33. Compass Sel.	#1
34. Flt Dir. Command Sel.	#1
35. Mach. Warn.	Ck
36. KLFLS	Ck
37. Yaw Dampers	On
38. Speed Brake	Dwn
39. Stab. Trim	Ck and On
40. Autopilot and Flt Dir.	Ck and Off
41. Trim Tabs	Ck

When Gear Doors and Flaps Clear

42. Rudder Power	Ck and On
43. Grnd Interconnect	Closed and Ck

Flight Engineer

12. Fuel Dump Panel	Ck
13. Circuit Breakers	Ck
14. Oxygen Mask and Reg.	Ck and Set
15. Oxygen System	On and Ck
16. Air Cond. Panel	Ck and Set
17. Galley Pwr	On
18. Generator Controls	Ck
19. Anti-Ice Auto Trip	Ck and Reset
20. Overheat Warn	Ck
21. DC Power	Ck
22. Eng. Fuel Valves	Open
23. Fuel Heat	Ck and Off
24. System *A* Pumps	On
25. Oil Cooler	Norm

After Starting

Pilots

1. Start Levers	Idle Det
2 Beacon Lts.	On
3. Start Sws.	Off
4. Flt. Controls	Ck

Flight Engineer

1. Electrical Power	Ck
2. Essential Power	Ck
3. Boost Pumps	On

4. Crossfeed Valves	Closed
5. Sys. B Pumps	Both On
6. Hyd. Sys. Pressure	Ck
7. No. 2 Bleed Valves	Closed
8. Pack Valves	Open
9. Door Warning Lts.	Ck

The 721-31 checklist contains eight specific checks divided among two types of operators and two operating situations. Such adherence to detail, coupled with a standard use of abbreviations, provides a meaningful instrument for assuring complete and adequate direction to the operator for normal or emergency actions. If more checklists were utilized in industry and government, there would be fewer interpretive errors which so often plague otherwise clearly written procedures.

The following checklist illustrates one of the typical aircraft "Emergency Guide" instructions. This one covers single engine failure, flap alignment, fuel hydraulic, and other emergency situations in a 727-31 jet aircraft.

Note that the following list is not labeled *checklist,* but instead is called a *guide* by the airline and F.A.A. approving agency. This is primarily done to indicate that circumstances could occur which would require a change in the engine shut procedure, for example. The term *checklist* is kept for those operations which almost certainly are to be done in the normal way indicated by the procedure.

727-31 Other Emergencies Guide

Single Engine Procedure

Approach

1. Reduce weight as much as practical.
2. Complete preliminary checklist.

*3. Review missed approach procedure.

4. Set bug to — 30 KTS.
5. Entering base leg, flaps 2%, 190 KTS. Min.
6. On final gear down, flaps 5%, and reduce to bug KTS.
7. Complete final checklist.
8. Commit is approx. 800 ft.

***Missed Approach**

1. Apply takeoff thrust, flaps 2%, gear up.
2. Accept altitude loss to accelerate to 180 KTS.
3. At 180 KTS, retract flaps and continue acceleration to 200 KTS.
4. Clear at KTS.

Flapp Asymmetry

1. Complete preliminary checklist.
2. If one device not extended:
   a. Set bug to — 15 KTS.
   b. Use normal landing technique.
3. If trailing edge flaps shut off by asymmetry system:
   a. Do not override.
   b. Set bug to — 15 KTS.
   c. Extend operable flaps.
   d. Use normal approach slot, reduce to bug KTS. on final.
4. If two devices not extended:
   a. retract flaps.
   b. reduce weight as much as practical.
   b. reduce weight as much as practical.
   c. Set bug to — 50 KTS.
   d. Gear down.
   e. Complete final checklist.
   f. Minimum maneuvering, bug — 20 KTS.
   g. Use normal approach slot, reduce to bug KTS. on final.

Minimum Fuel Procedure

1. All boost pumps on .... On.
2. All crossfeeds open .... Open.
3. On missed approach, avoid sustained nose up attitudes beyond 10%and steep turns.

Manual Reversion Procedure

1. Complete preliminary checklist.
2. Set Bu to V_{ref} — 15 KTS.
3. Plan for manual gear extension.
4. On downwind leg, extend 15% flaps.
5. Use normal approach slot, reduce to bug — 5 KTS. on final.

Gear Lever Locked

1. If greater emergency exists, override lever lock and retract gear.
2. Nose steering wheel neutral, OK.. OK.
3. If wheel does not remain in neutral, do not retract gear.
4. With wheel neutral and lever locked:
   a. Both safety relay C.Bs (P6) pull, pull.
   b. Override lever lock and retract gear.
5. Just after touchdown, reset safety relay C.Bs.

Hydraulic Pump Failure

1. Failed pump off .... Off.
2. If system pressure lost, yaw damper off .... Off.
3. If no fluid loss, other pump on .... On.
4. When system pressure normal, yaw damper on .... On.

Hydraulic Fluid Loss

1. Affected system pumps off .... Off.
2. If A system and loss continues, pump switches to depr .... Depr.
3. Yaw damper off .... Off.
4. If loss stops, return pumps one at a time to depr. and check quantity.
5. If unable to restore both yaw dampers, descend to 26,000 feet.
6. See 03.12 for system restoration procedures

Hydraulic Fluid Overheat

1. After gear and flaps are up, affected system pumps off.
2. If A system, pump switches to depr.
3. Yaw damper off .... Off.
4. Restore systems on approach.
5. Yaw damper on .... On.

The critics of checklists consider them "after-the-fact," "too general," "defensive," or that "nobody will use them anyway." These comments can be true if the associated management disciplines and policies are not established and followed. If an X-ray machine operator does not follow the procedures for storing and handling the film, or shortcuts the machine warm-up period specified in the operating manual, then there is little reason to believe the operator will follow a checklist for the final sequence and settings of high voltage, exposure time, radiation protection, etc.

However, with management encouragement, checklists can serve the purpose of reducing the tendency of operators, designers, technicians, mechanics, etc., to institute personal shortcuts, the majority of which may cause any of the four R's: Risk, Rework, Repair, and (w)Recks.

A checklist itself will not cure all the ills of a new design, or guarantee the perfect initial setup of a $100,000 numerically controlled machine, but following a checklist will assure (within the limits of human error or haste) that predetermined parameters are considered. A checklist is particularly applicable where the sequence of events is important.

The following are classic examples of where checklists saved the day, and where they caused problems.

It was an error in a NASA checklist that caused Apollo 10 (with astronauts Strafford, Cernan, and Young) to roll nearly out of control on May 23, 1969. During descent to an altitude of 9.4 miles of the lunar surface, a switch was moved as prescribed in the operating checklist. However, the checklist was in error and the fault was promptly corrected by a well trained crew. Without checklists, however, the fantastically complex Apollo missions could never have been successfully accomplished.

Checklists are the essential control in the start-up and shutdown of nuclear power plants (both military and civilian), aircraft engine operation, and missile countdown procedures. Checklists are used in transmission rebuilding manuals, in some carburetor repair kit instructions, and even in a number of published cookbooks. Checklists are utilized in hundreds, perhaps thousands of process industries —from electroplating shops to epoxy companies to metal alloy producers.

Troubleshooting checklists, sometimes in matrix form, have been used in the electronics industry. Howard W. Sams, Inc. publishes *Photofacts* for TV, radio, hi-fi, etc., and pioneered the standardized schematic format and troubleshooting checklists. These became the backbone of technical reference for the nation's nearly 100,000 TV service shops.

The author has seen one sentence checklists in outboard-motor repair manuals, lists as simple as three-step checklist questions in communication receiver/transmitter operating instructions, and five-item checklists for office typewriters with replaceable type features.

A checklist is not to be confused with directions for assembling a Christmas toy, or with a cake recipe, or even with the step-by-step procedure for completing an IRS 1040 Form.

This text does not consider basic instructions, procedures, directions, programs, etc., as a checklist. In industrial applications, this book considers checklists as primarily supplementary documents used to recheck the original activity. This concept does not limit them just to inspection or test uses, but rather is premised on individual assemblers, operators, designers, etc., utilizing this technique at selected points.

The public accepts the use of checklists on most military and civilian firearms ranges—All ready on the right; All ready on the left; All ready on the firing line; Fire.

In a tour the author took through the Oyster Creek Nuclear Electric Generating Station (the first operational atomic plant in New Jersey), it was reassuring to note that detailed checklists were extensively used by the operators. They are strictly followed during start-up and shutdown, and during the hundreds of associated procedures, such as withdrawing improperly functioning fuel rods. If this 640,000 KVA plant were within three miles of your home, the author is sure you would want to know that procedures approved by the Atomic Energy Commission were being followed.

The reader may reply that in these examples safety considerations were the overriding factors which justify the use of checklists. That is true.

The author is not recommending a checklist as a sure cure for all the ills of industry, labor, management, government, design test, etc. What the author proposes is selected use of checklists for situations where the opportunity for error can be significantly reduced by a review of factors which history proves cause repetitive errors.

Certainly, a checklist per se is not needed to start a car, mix a cocktail, buy stock, or pour a concrete slab (though the author has seen satirical checklists written on each subject).

Let us establish the need, which varies of course for individuals, for a checklist in the following three examples.

1. When you close up your house for a vacation, most people put in practice (or wish they had) an unwritten checklist (some even jot the items on a list), such as:
   a. Is main water valve off?
   b. Is thermostat off?
   c. Are windows locked?
   d. Are doors locked?
   e. Are milk, cleaners, papers, etc., services stopped?
   f. What lights are to be left on?
   g. Is garage locked?
   h. Are neighbors alerted to pick up mail?
   i. Have police been notified?
   j. Who will cut grass?

The aim in a vacation list is to spend minutes to save dollars and reduce financial risk or worry.

2. Before taking your car to the state inspection (or asking your wife to), the majority of car owners check the obvious items that could reject the car, such as:

a. Do lights work: park, stop, license plate, headlights (hi and low beams)?
b. Do blinker signals work?
c. Do wipers work?
d. Is level of break pedal proper? Is it spongy?
e. Does emergency brake work?
f. Does horn work?
g. Is license plate(s) legible?
h. Is any glass cracked? Is all glass clear?
i. Do door latches work properly?
j. Are tires OK?
k. Is registration card available?
l. Is there noticeable kingpin or tie-rod wear in the front end?

The approach in a car inspection checklist is to make a routine check to avoid the inconvenience of possibly having to waste hours and/or incur unnecessary repair charges because an obvious item was not checked prior to the official inspection. How many times has the reader wished he had made such a check *before* going to the inspection station?

3. In selecting a new home, a sharp buyer should review (using pencil and paper) a checklist in a tradeoff of costs vs. facilities, such as:

Costs
a. Taxes
b. Interest and points
c. Mortgage (FHA, VA, Conventional)
d. Hidden charges
e. Closing costs
f. Conditions in agreement of sale
g. Insurance
h. Liens
i. Assessments
j. Moving expenses
k. Dues

Facilities
a. Schools
b. Utilities (gas, electric, water, sewers)
c. Recreation facilities
d. Shopping
e. Roads, sidewalks

f. Churches
g. Transportation
h. Fire and police protection
i. Traffic
j. Zoning

A real expert comparison shopper or professional broker, builder, banker, etc., might expand the first two cost items as follows:

Costs

Taxes

a. What is tax rate per $100 of assessed valuation?
b. How much were taxes last year on a similar house?
c. What is the ratio of appraised value to true selling price?
d. Are any assessments forthcoming?
e. How many schools, libraries, etc. are planned to be appropriated in the next few years?
f. When was last municipal upgrading of tax appraisals?
g. How much more are taxes for a corner lot?
h. Will tax assessor check inside of house?
i. Are school, sewer, or garbage taxes extra?
j. Is there a personal property tax?
k. What portion of taxes are paid in advance to mortgage company?
l. Do you qualify for a veterans tax exemption?
m. Is there an income tax?

Interest

a. What percent is rate?
b. Are points required?
c. Are there any other special charges?
d. What is policy on late monthly charges?
e. Is there a premium for mortgage prepayment?
f. Is life insurance coverage required?

Virtually all persons reading this page, who have bought a new house recently, will admit this expansion into a checklist of the considerations affecting a new home's tax and interest considerations identifies several factors which were a surprise to them in buying their last house. They will admit that a list such as this could have alerted them to a better negotiating position, and even saved them money, aggravation, and delays.

These examples are a lesson in the usefulness and practicality of good checklists for selected situations. *The checklist concept is a useful tool,* but it has not been effectively used by most professional people, by industry, or by the government. In these times of constantly rising costs, checklists can be a form of "memory feedback system" to *reduce errors,* expedite *better decisions,* provide *improved problem appraisal,* and to *reduce unnecessary expenses.*

Checklists can also provide:

1. Faster break-in of new employees
2. Easier upgrading of skills of older personnel
3. Changing the skills of salary and hourly people to meet changing requirements
4. Written reminders for critical or sensitive operations
5. A "check and balance" on important decisions
6. That "lessons learned" are considered in current decisions.

The Checklist Concept

1. Prepare checklists as working level guides where necessary to supplement other basic documents.
2. Keep them short, concise, flexible—and up-to-date.
3. Back up with pictures, models, and visual aids as required.
4. Integrate with operations performing the task.
5. Once you've got good checklists—use them.

Try The Checklist Concept—It Works.

Chapter 2

WHEN, WHO, AND HOW TO PREPARE AN EFFECTIVE CHECKLIST

Without engaging in a discussion of organizational structure and philosophy, the following premises are established for the preparation of checklists for industry and government. Realizing that checklists are not a cure-all, that they cannot be all inclusive, that they are only to be established in certain applications, and that they often supplement (not replace) other written requirements, the following guidelines are suggested:

(*Note:* The following might be called a checklist for checklists.)

1. A checklist should be as brief and concise as practical.
2. A checklist is to be tied by revision (where required) to changes in bulletins, drawings, standards, specifications, planning, etc.
3. Every checklist should bear the date and name of initiating person and approval(s).
4. Any checklist for use by manufacturing and/or inspection is to be coordinated with both groups.
5. Visual aids (photos, models, cut-a-way views, etc.) and accept/reject criteria can be part of a checklist.
6. Checklists which accompany a product (to the next station or as part of final shipment) should indicate (by final written stamp or signature) compliance to requirements up to that station.
7. The goal of every checklist is to provide a self-check by the individual, and thereby actually reduce duplicate inspections, error hazards, etc.

8. The principle of the "check and balance system" should be utilized in all checklists which perform a controlling function, i.e. the group establishing the checklist should always integrate the document with the performing group before it is instituted.
9. Any person preparing a checklist should prepare it in the work (cockpit) area with his hands touching the actual parts, levers, dials, etc.
10. Checklists should be updated as necessary to keep them accurate and effective.

When to Prepare a Checklist

When is a checklist needed? When is it best to mark up a visual aid photo instead of writing words? When is the optimum time to implement the checklist? When is it better to include the checklist items on a flow chart?

In *real life* research, development, and production work, the product design, configuration, costs, materials, processes, etc. are often being refined or undergoing changes while a high quality/reliability product is still required for on-time delivery to the customer.

Frequently in development and often in production the product improvement has to be incorporated with the day-to-day production control problems resulting from:

1. Wrong parts
2. Substitute parts
3. Late parts
4. Missing parts
5. Suspect parts
6. Defective parts
7. Reworked parts.

For the basic industrial categories below, there follows a discussion of recommended ground rules and examples when a checklist should be prepared:

1. Design
2. Processes
3. Inspection/Test
4. Procedural.

The detailed guidelines for the aforementioned areas are:

Design

1. The drawings should be complete. When they aren't, a checklist can be used in certain applications on the manufacturing

floor as a temporary substitute for the missing information in order to move hardware. Checklist 35 in this volume, "Multicoder/Multiplexer Checklist," is an excellent example of a design-oriented checklist which can highlight the factors necessary to proceed with on a rush design.

2. The specification (performance, design, test, inspection, etc.) and processes procedures should be prepared or modified prior to their need in the operating areas. In the absence of a revised document, a marked-up checklist can often suffice as an adequate temporary process control.

The question in each of these areas is when is it better to expedite the final release of a drawing (or specification), or press for the needed change to a drawing, specification or procedure, and when is it more effective to initiate a quickie (often on-the-spot) checklist.

Checklist 23, "Cable, Harness, and Connector Checklist," was initiated in 1967 when connector problems seemed to be plaguing one major aerospace company at every turn. The following year when a rash of coaxial connector problems developed, Checklist 24, "Coaxial Cable/Connector Checklist," was written.

Processes

An example that shows how a timely release of a checklist can be used to reduce controversy and expedite current hardware is the need General Electric had to establish criteria for microelectronic processes which occurred on a dozen suppliers of devices for its Minuteman III program in 1967. Checklist 44, "Microelectronics Process/Quality Checklist," was then written and released four months prior to the release of the first government specification (MIL-STD-883) on microelectronics. This identified more than 250 process points where decisions had to be made and understood by the various design, manufacturing, process, and quality personnel.

Of course, the degree of detail of bench top process/quality checklists should be brief and very specific to avoid an interpretation problem similar to those in drawings or specs which may have caused a need for a checklist in the first place.

In Checklist 36, "Printed Circuit Board Checklist," Item 12 under the section "Shop Processing" asks the question: "Do boards show any evidence of blistering, delamination measling, or crazing after dipping in 60-40 composition solder in accordance with applicable NEMA or IPC or government standards?"

This item actually covers three phases in a single interrogative sentence.

The first part asks if the reader has considered the visual accep-

Figure 2.

tance of several common discrepancies. The second portion assumes "solder dipping in 60-40 composition." The reader may utilize wave soldering in a 63-37 composition—and that may be acceptable. The point is that reference is made to one specific method and exact alloy.

The final point is that the subject item is to be considered in light of the applicable referenced specifications which are listed.

By implication, the reader can assume that is an approved method, but since a checklist cannot be all inclusive, the process can be performed by other acceptable processes. The point is that the process used is to be considered with respect to that listed in the checklist.

A point of clarification is appropriate here. Some companies refer to their detail inspection planning instruction sheets as inspection checklists. Such a use of the word "*checklist*" is not intended by this volume. A checklist, as recommended here, is intended as a *tool* to supplement (not replace) *in selected applications* the procedures, drawings, specifications, instructions, etc.

Inspection[1] Test

Proper crosswire or spot welding requires perhaps a dozen parameters (pressure, voltage, current, probe tips, etc.) to be in the proper condition. A checklist on the subject in Vol. VIII can aid in setting up, running inspection and test coupons, sampling the product run, and troubleshooting the installation. It significantly aids in providing important guidance to designers and planners who only infrequently witness such operations.

Examining this checklist, one can see the depth of detail in a good inspection/test checklist.

The "Printed Circuit Board Checklist," while heavy in its design and process considerations, provides a strong insight into the many accept/reject criteria points which need to be determined to assure adequate inspection and test of this product. It was prepared to force a discussion of such details which were being masked in a shield of indecision in a number of procurement specifications.

Procedural

The "Laser Checklist" (34) is an example of a checklist which endeavors to cover all the procurement considerations for a relatively new type of product. While it is supplemented by detail drawings and specification sheets, in itself it provides both in-plant and customer procedural direction as to the decisions that have to be made in selecting a device like a laser.

Who Should Prepare a Checklist

Who should write a checklist? The answer is anyone—anyone who knows the details of the situation for which the checklist is intended. For example:

1. A chief engineer should have his standards engineer add to the ball bearing checklist in his design manual the two new factors learned from the failure analysis on the new high speed air driven turbine with oil mist lubrication of its ball bearings.
2. A foreman should write a few sentences or draw a sketch to clarify the routing of a substitute part's leads in a high frequency circuit or for the special check of tolerances on a tricky assembly sequence. A simple sketch or a few words can be considered a checklist, but they should be displayed in the immediate operator's working view.
3. A keypunch supervisor can obtain an effective reduction in errors and improve efficiency by displaying a list of items to check before putting cards in the verifier or sorter.
4. A sketch should be drawn to clarify the magnified view of an allowable epoxy separation as compared to a rejectable crack on a microelectronic device. Engineering, manufacturing, and quality control should agree to the accept/reject criteria. Such a checklist can be implemented almost immediately, long before the standards (if utilized) are drawn up, or revised, then approved, released, and implemented.
5. A reliability or quality configuration control reviewer should have a checklist available (perhaps to more than one level of detail) to demonstrate the controls at various points in the software flow.

Every checklist should bear the signature of the person initiating it (and the person accepting it, if other than the originator) and the date released. A revision letter is usually applied on more formal checklists in order to make it a controlled document.

How to Prepare a Checklist

The format of a checklist can vary, as it does among those in this manual, but all should bear a single commonality.

Checklists Should Be

1. brief and concise.
2. The words need not form a grammatical sentence.

3. It is optional whether the point is raised in a question-type format or in a statement of fact.
4. The items in the checklist may be grouped by sequence or category, as applicable.

Checklists Should Not Be

1. written in essay format.
2. They should not list repetitive items unless a duplicate check is ncccssary.
3. Checklists should not be general in nature.

Checklists may, on occasion, be general, but to really be effective they should be specific. How specific? That depends on many factors, such as whether the checklist covers a product, process, or a philosophical approach to management problem solving.

This volume contains checklists with detail varying from the moderately general "Electrical/Electronic Reliability Checklist" (29), to the explicit "Transponder/Beacon Design Checklist" (42), to the very detailed and pacesetting "Microelectronics Process/Quality Checklist."

The accepted criticism of checklists to date, in industry, has been that they are often written by persons in isolated standards or reliability groups who did not have to "sell them" to the user, and who were isolated from the responsibility of assuring they were followed in actual use.

Chapter 3

ELECTRICAL/ELECTRONIC CHECKLISTS

23. Cable, Harness, and Connector Checklist

Andy Cutler
Process Control Specialist
General Electric Company

No other single product line causes more consternation in receiving inspection, assembly, stockrooms, or test than *connectors*—as used in cables, harnesses, racks, and black boxes. NASA and the military have repeatedly stated that the number one cause of their problems is connectors.

More than *60* companies are manufacturing an estimated *100,000* combinations of connector types, sizes, alignments, and materials. This checklist has been prepared to assure the myriad of factors affecting cable, harnesses, and connectors are considered in all phases of design, manufacture, inspection, test, and use.

It has been successfully used by General Electric's Re-entry and Environmental Systems Division which procures more than 50,000 connectors a year. The author, who has worked full time for more than ten years on the problems described in this checklist, also led the effort that caught a defect which, in 1967, required one connector supplier to institute a nationwide call-back of tens of thousands of defective connectors.

The emphasis on this area of electrical/electronic problems can be indicated by the fact that three checklists in this volume are associated with cables, harnesses, and connectors. These are: "Cable, Harness, and Connector Checklist" (23); "Coaxial Cable/Connector Checklist" (24); and "Flexible Cable and Circuitry Checklist" (33).

Design and Mock-up

Cable and Harness

1. When do you specify resilient potting? Hard potting? Clear potting? Epoxy? Room temperature vs. oven cure?
2. Is the minimum wire insulation thickness adequate? Will cold flow deform insulation?
3. At what temperature will insulations:
    a. outgas
    b. become brittle
    c. become toxic?
4. Are metallicor plastic adapters needed in breakouts?
5. Is design such that cable or harness is flexible? Has obsolete lacing tape been replaced by spiral wrap, tape wrap, or shrink tubing?

6. Should insulation be radiation hardened? Is it combustible?
7. On highly flexed wires, is oxygen-free copper wire specified?
8. Is a three-dimensional fixture used to *build and test all complex* flight cables and harnesses? Is a two-dimensional fixture used on others?
9. Is a twist in cable wires specified for all points of flexure?
10. Is tubular or wrapped shielding to be used?

11. Are point-to-point wiring call-outs provided by a drawing or interconnection list?
12. Is an effective system provided to minimize cable and harness changes, and adequately implement required changes? A control configuration?
13. Is a conductive adhesive compound required for adequate outer shielding ground at megacycle and gigacycle frequencies? See "EMI/RFI Checklist" (32).
14. Is RTV (or equivalent) coating needed over RTV taping for protection and moisture barrier?
15. Is shrink tubing with transitions required for main trunks and breakouts for protection?

16. Are cable assembly breakouts dimensioned within the specified one-quarter inch tolerance?
17. Are overall dimensions maintained from the front face of each connector?
18. Are connectors mounted horizontally for minimum moisture entry?
19. Are overall dimensions maintained from the front face of each connector?

For hermetically sealed connectors, see "Hermetic Sealing and Leakage Rate Checklist" in Vol. IV.

Connectors and Contacts

1. Are female connector contacts designed with splits which permit 360° engagement of surface area in bearing when mated with male contact?
2. What is inspection plan for bent or misaligned pins? What is criteria for accept/rejection? When are bent pins to be replaced rather than rebent?
3. Are gauges used for check for pin damage, bending?
4. Are solder-well depths sample inspected to be sure they are not overdrilled (which can cause telescoping and breakage of pin)?
5. Do applicable connectors meet requirements of:
   a. MIL-C-26482
   b. MIL-C-26500
   c. MIL-C-27599
   d. MIL-C-38999
   e. MIL-C-39029
   f. MIL-C-55302?
6. Can insertion of pins cause socket to fail in shear or compression?
7. What steps are taken to protect connector couplings from surface damage when attempting to loosen tight connector couplings?
8. Have the JT connector solder well and swaged hood on socket contacts been gauge checked?
9. Are JT-RE crimp-type retention clips properly located? Is there any evidence of sealer seepage into the contact cavity?
10. Have Deutsch crimp-type (STK) series connector contacts received gauge check for Split *C* locking device control?

11. Can you assure that the hand crimping tool bears evidence of being recently calibrated? What is the recalibration cycle? Were sample crimps made?
12. On crimp-type contact terminals and spade lugs, are crimped indents uniform? Is crimp barrel excessively out of round? Tarnished? Is gold plating flaking or fractured?
13. Is the insulation clearance on crimp-type contacts between .010-.050 where the contact barrel does not contain provisions for crimping the insulations? In JT and LJT crimp type socket contacts, are only the four segmented brass barrels used since contact resistance can vary in singly dimpled barrels?
14. Where crimping of the conductor and insulation is applicable (by nature of contact or terminal design), are they examined to assure that the center conductor reaches the visible inspection hole, and that the insulation is secure in the crimped barrel portion provided for this application?

    Note: *a.* Do unused connector cavities contain contacts and sealing plugs?

    b. Are sealing plugs (plug end seal) on JT and LJT crimp insertable connectors inserted knob end first?
15. Are visual checks performed for evidence of punctures or indentations on connector insert fact to photographic illustrations shown in standards?
16. Are the contacts secure and uniform in height with reference to the insert and shell as referenced in the drawing, including misalignment and radial displacement?
17. Is connector coupling engagement and mate/demate possible without the aid of special tools? If required, is there room for tools?
18. Are visual checks made for evidence of damaged ground fingers?
19. Since there was no manufacturer direction up to November 1969 as to which end of a plug end seal was to be inserted into the rear insert (grommet) on JT and LJT connectors, these points should be checked. The proper way is to insert the flared end into the rear insert.
20. Is Sunbank rear end hardware specified on critical EMI/RFI designs? Where used, is connector adapter torqued to a sufficient value that when Sunbank clamp is unscrewed, the adapter will not back out too? (Applied to two-piece construction such as on MS 27468.)

Shield and Soldering

1. Will silver coating on wire cause too great an affinity to solder and require the use of anti-wicking tools?
2. Are proper ferrules used to terminate shielding at all ends?
3. Is inner shielding to be terminated inside or outside of connector potting? Assure that shields are trimmed as close as possible to the outer insulation and sleeve where they are not electrically common.
4. Have in-process soldered connections been examined, say at 25, 50, 75 and 100 per cent points in the total connector soldering compliment?
5. Are good wetting, adherence, visible solder well and wire configuration (outline) cleanliness, excess solder, and wicking proper?

6. Have all shield terminations and dress been checked for the following:
   a. Dimensional to drawing and specification requirements?
   b. Solder ferrule application and jumper leads to approved sample/photographic illustration?
   c. That the shield terminations are proportionately staggered when solder ferrules are utilized?
   d. Floating shields covered with shrink sleeving and embedded in potting when individual shielded wires are terminated?
   e. On multiple shields with a common tie(s), is shrink sleeving required and/or embedment in potting required?
   f. That the shields are trimmed as close as possible to the outer dielectric insulation?
   g. The application of heat shrink sleeving under shield terminations at solder ferrules and EMI "daisy chain"?
7. Can you assure that solder ferrules when used for splicing or shield terminations are not installed on the radius portion (area) of cable?
8. On EMI hardware applications: (See also "EMI/RFI Checklist," 32.)
   a. Are the individual shielded wires jumpered and terminated by soldering to the metallic potting shield (boot)?

b. Is the spiral wrapped shield applied properly and soldered to the potting shield (boot) where applicable? Where a tubelike shield is employed, do the same conditions apply as for spiral? In both cases, is insulating coating applied over the shields? Does this operation check the same points as item number three in "Potting and Associated Operations" in the section below?

c. Can you assure that the adapter and potting shield have been adequately secured electrically and mechanically to the connector shell?

d. Has soldering of adapter and potting shield been checked?

e. Has continuity of shield terminations from connector shells to connector shells throughout the cable assembly been checked, including the RFI Sunbank overbraid shield terminations and ground joint adapters, where applicable?

9. Are adapter pins properly crimped and bottomed in solder well?

See the checklist on soldering in Vol. VIII.

Wire Dress and Identification

1. Are there any "rats' nests" of dense, disordered arrays of wires?
2. Are wiring channels overcrowded?
3. Is lead dress specified at breakouts where necessary to provide "neat" wiring? Is crossover of leads allowed? How much?
4. Are all requirements of MIL-E-4148, MIL-E-5400, MIL-W-8160, and MIL-E-8189 met if applicable?
5. Are harness ties properly applied and spaced?
6. Are proper methods and length of air frame ground terminations used?
7. Is insulation clearance between dielectric and soldered connections as follows:

a. Is maximum clearance equal to the outside diameter of the insulation plus 1/32 inch applied to wire where the outside diameter is greater than 1/32? Is the minimum insulation clearance allowable in this case 1/32 inch?

b. Is maximum insulation clearance of 1/32 inch applied

to lead wire where the outside diameter is less than 1/32 inch?

8. Is the minimum clearance in case *b* the outside diameter of the wire?
9. Does application of identification markers to each individual lead comply with MIL-W-8160, if applicable? Are they within two inches of the prescribed connector potting mold?
10. Have the identification markers been properly applied per parts list call-out? Do they contain the following information: Lot or S/N, drawing, revision, cable nomenclature, (example W2E)?
11. Can you assure that the solder ferrules when used for splicing or shield terminations are not installed on the radius portion (area) of cable?
12. Have provisions been made to detail the application of shrink sleeving under shield termination and solder ferrules?

Potting and Associated Operations

1. Are adapter ring and potting molds bottomed?
2. Are proper size and angle potting cups used where required?
3. Is potting "growing" inside connector due to improper curing? Do tightening rings turn harder after being subjected to high temperature?
4. Are you certain free floating pins are not restricted by potting of leads?
5. Are you sure all surfaces requiring priming prior to potting actually get primed? Is it written in planning?
6. Are detailed sketches and/or method sheets provided to back-up planning to assure proper application of potting on connectors?
7. If Amphenol right angle potting molds are used on Bendix pigmy connectors, are special instructions provided?
8. What check is made to prevent use of oversized potting molds on connectors?
9. Shouldn't cables used in acceptance test areas be potted?
10. Are connectors designed for potting actually potted except where either strain relief is provided or moisture/dust is not a problem?

11. Do all Amphenol and Bendix pigmy-type connector coupling rings have a maximum of 1/32 inch axial play and are they free to "rotate" at time of inspection?

12. Has potting been checked for the following:
    a. Correct potting mold and size?
    b. Alignment of potting mold with respect to connector shell and keyway orientation?
    c. That the individual wires are surrounded with potting?
    d. Evidence of air bubbles adjacent to or between soldered connections and/or leads?
    e. Potting contamination/discoloration?
    f. Monitor elevated temperature cure cycle for potting?
    g. Is potting viscous enough that it doesn't run past mold?
    h. To assure a primer is used for proper polymerization if base material is inert such as zinc, cadmium, or gold?

13. Are the in-flight disconnects potted while mated to the precision fixture applicable for control of contact alignment and uniformity? Does identification comply with MIL-W-8160, if applicable?

14. Is identification paint, EMI conductive paint, liquid sealants, or potting in any way restricting normal mate/demate by increasing torque? Are slivers generated by excessively tight fits on connectors?

15. In pressing potting cup down snug:
    a. Is potting cup factory installed or installed after soldering?
    b. Is potting cup always pushed back during soldering?
    c. Are torque values known for mating? Are they different from demate torques?
    d. Are high torque JT connectors used in any near inaccessible or blind locations?
    e. On a JT connector with more than twenty-four pins a tightening tool is needed. Is there room for tool?
    f. Have finished connectors been checked for proper axial movement against wave washer?

16. Cable Coating—Has in-process been checked to insure that the breakouts are properly and adequately taped? Is the RTV (or equivalent) taped properly and evenly applied?

17. Was priming accomplished and was the RTV (or equivalent) coating uniformly applied?
18. Are all lacing ties and tape applied in a manner to prevent stress on connector contacts and leads where crimp insertable connectors are utilized? At what maximum distances for what size bundles?
19. Are the elevated cure temperatures monitored and referenced on planning? (See "Potting, Epoxy, and Conformal Coating Checklist," in Vol. VIII.)
20. Are strain relief cable clamps installed using vendors recommended holding and tightening tools?

Inspection and Test

1. Is a connector mate/demate log needed to reduce problems in testing areas?
2. Are "connector savers" or "socket savers" called out for points of frequent test connecting?
3. Are all hipot (high potential) and insulation resistance tests performed according to procedure 303 and 304, respectively, of MIL-STD-202?
4. Do your people know the real difference between what hipot and insulation resistance tests are intended to check?
5. Is a maximum allowable leakage current (in microamps) specified for hipot tests?
6. Upon completion of fabrication or repair of test cables, are they resubmitted to inspection/test?
7. Does inspection check the cable per standard requirements and/or to the detailed wiring diagram/schematic/point-to-point drawing?
8. Is each test cable used for system, component, or module testing identified?
9. Is it necessary to record weight? On flight cables?
10. Is each test cable used for harness testing hipotted and insulation resistance tested?
11. Are all cables inspected by the user (inspection/tester)upon issuance from the storage areas and inspected upon return to the storage area?
12. Which test cables should be handled as "prime" equipment?

13. Are all cables and harnesses handled via supporting fixtures or protective containers?
14. Do test cables have their connectors protected at all times (when not in use) with protective covers? If not, how are they protected?
15. Are all test cables, when in use, adequately supported to prevent undue stresses at individual connectors?

16. *Have you stated you'll give a written warning notice to anyone who leaves a cable or harness on the floor?*
17. Has the cable assembly fixture been approved in accordance with the tool control card and released drawing configuration?
18. Is there evidence of previous hardware acceptance for all in-process operations?
19. Is it certain that free floating pins are not restricted from movement by potting of leads?
20. Has provision been made to lot sample inspect cable connectors and terminations by use of X-ray?

21. Are proper ferrules used to terminate shielding at all ends?
22. Is verification of operator certification required for wiring, soldering, crimping, and potting?
23. Are "in-cable" splices of wires used with heat shrinkable solder ferrules (by thermogun or zap gun) dimensionally inspected to illustrations? Are they allowed by contract?
24. Are nicked wire strands checked for missing stands, unravelling of strands, for normal lay, evidence for bare copper, and damaged/nicked shields?
25. Are visual inspections made for nicked, bruised, damaged insulation?

26. Are visual inspections made for burnt, discolored insulation (other than superficial discoloration caused by normal thermal stripper usage)?
27. Has continuity been performed in accordance with wiring interconnecting drawing? Are adapter mating test cables utilized? Is the continuity performed with a continuity test voltage, current limited to 50 ma, on a half ohm maximum resistance device (tester)?
28. Has a high-potential test of 500, 750, or 1,000 (−0 + 75) volts AC RMS 60 cycles, held for a minimum of two seconds,

been made between *each conductor and its shield?* Between *each conductor and the connector shell?* What is allowable breakdown current in microamperes?

29. Has an insulation resistance test been made between *each conductor and every other conductor and its shield, and between each conductor and shell* at 50 or 500 D.C.?
30. On multiple shield cables tied to a single conductor pin, the voltage applied for test may be substantially reduced because of capacitive effects. Has provision been made to correct for this?
31. Is grommet support provided between strain relief clamp and harness bundle? Are sleeves and shields adequately secured within strain relief clamp?
32. Does in-process inspection insure breakouts are properly located and prepared?

Packaging and Storage

1. Upon acceptance, who packs and submits the cable to the applicable storage crib?
2. Are all cables not in use returned promptly to the storage area?
3. If, upon return to the storage area any defects exist in a cable or harness, are such defects recorded and communicated to the storage area attendent for processing a replacement/repair request? Or do they lay hidden until needed again?
4. Are cables and harnesses packaged in their normal lay (if possible) in sealed plastic bags, and secured in a corrugated box? Do they contain the travel tag, mate/demate log?

Special Items

1. Will mating connectors or connector covers be allowed to be attached during humidity, rain, or salt spray tests? If so, it makes the test easier to pass.
2. Are holes put in "black boxes" to prevent moisture-laden air from breathing through unsealed connectors?
3. Have all operating personnel been carefully orientated with regard to:
   a. planning package contents, visual aids, tools
   b. standard workmanship requirements
   c. quality control levels and accept/reject criteria

   d. typical detailed (exploded view where necessary) drawings
   e. interconnection drawings and notes
   f. cable configuration drawing including parts list?
4. Have proceedings of the National Electronic Packaging and Production Conference been reviewed for latest methods?

REFERENCE

"Electrical/Electronic Inter-Connection and Connector Techniques," by R. Lawrence, Senior Vice President, Deutsch, is recommended.

24. Coaxial Cable/Connector Checklist

E. F. Coulson
Component Parts Specialist
General Electric Company

Within the realm of cables, harnesses, and connectors is the separate group of coaxial problems. Whether they be associated with VHF, or UHF communications, of the newest radar and stripline circuits, coaxial cables and connectors.

This checklist is a supplement to Checklist 23 and applies to both rigid and flexible coaxial cables.

Design

1. Since excessive flexing of the coax cable assembly can adversely affect the inner and outer conductors, has every practical precaution been taken to protect coaxial cable from stresses of flexing?
2. Do point-to-point dimensions meet the prescribed tolerance to assure VSWR requirements and control of insertion loss?
3. Are electrical tests performed as follows: per MIL-STD-202 or other applicable spec?
   a. — continuity test with limited current
   b. — high-potential dielectrical tests with maximum current.
4. Are contact withdrawal and engagement forces specified to assure good mating characteristics?
5. Is the contact properly secured to the center conductor (coax) either by solder, or crimp?

6. Is there a control for solder flux and solvent cleaning alcohol applications to prevent discoloration and corrosion on shield-

ing and center conductor which will effect VSWR, insertion losses, and reliability?

7. Are captive type contacts used, instead of noncaptive, for added handling protection?
8. Is a captive contact type connector used with RG142B/U coax wire?
9. Is axial movement of the center conductor, which occurs with flexing, under control? How do you know?
10. Is maximum play of center pin contact within limits?
11. Have "push coupled" coax connector engagement and withdrawal forces been checked?
12. Has a gentle controlled application of an axial rotating motion been made to the cable while holding connector to assure that the shielding is adequately secured?
13. Is connector the type that cannot be disassembled without destroying coupling nut?
14. What is upper frequency limit? Is it at or under the GHZ where the first higher order mode may propagate?
15. Are applicable parts of MIL-C-39012 applied?
16. Is damage point or breaking torque in assembly known? At what point will VSWR be affected?
17. Are connectors mounted horizontally for minimum moisture entry or are they in a moisture protected area?

Assembly

1. Is special vendor assembly tooling referenced on drawing?
2. Does planning reference special in-house assembly/tools?
3. Does planning reference contact dimensions from the same datum surface? For both male connector contact and female connector contact? Are these dimensions referenced with respect to the dielectric for both the male and female contacts?
4. Does planning reference removal of the preformed cable from assembly fixture and securing to a sturdy support which contains the drawn configuration? Is a support with the secured cable then put in a protective package?
5. Is preparation of coax cable planning for stripping of its jacket and dielectric adequate? Against damage, nicks? For pretinning?

6. Is shield dress and foldback adequate? Against damage, nicks?
7. Does the loosely assembled connector contain all the component parts, and are they arranged in accordance with cable assembly instructions? When shield termination requires soldering, is shield termination ring tinned?
8. Is packaging secured to card and sealed in plastic bag?
9. Are all possible steps taken to avoid excessive handling?
10. Are three-inch dimensional fixtures and packaging used where specific configurations are required?

11. Can the gland or clamp nut turn or swivel and damage the shield?
12. Where applicable, soldering of center conductor and contact should be done after the cable is clamped to prevent torque, tensile, compressive stress when assembling the securing hardware. Is it?
13. Assembly tooling specified by the respective connector manufacture should always be procured and utilized? Is it?
14. Since the connector component parts normally are received loose in a package, experience has dictated that the *loosely assembled connector parts in their respective order of sequence* must be closely examined to insure that the shield locking mechanism is in the proper orientation. This has proven to be the most critical check point, and it is stressed in the planning package. Has it been done?
15. Is proper handling heavily stressed by illustrations and education?

Inspection Test

1. Are all coaxial cables packed securely in closed protective containers?
2. Is the support secured cable flexed at all junctions during VSWR' tests? Are all possible tests performed while cable is secured to the support? Does the package remain intact until ready for installation at final assembly?
3. Are torque wrench call-outs for proper tightening of coupling nuts when mating to counterpart receptacles specified? Is there an overall maximum limit of specified torque displayed for each applicable size and type coupling nut? Are couplings torque proof tested?
4. Is there evidence of previous material acceptance? Is plating tarnished?

5. Are solder-type connections inspected to assure that the soldered contacts are secured and snug against the dielectric?

6. Are crimped-type connections inspected to insure that the contact and shield are properly crimped utilizing calibrated hand crimping tools? Is the contact snug against the dielectric?
7. Are the connectors properly protected with protective devices (dust covers, holding fixtures, etc.)?
8. Are the male and female contacts seated within the specified dimensions with respect to the datum surface?
9. Is the potting properly applied to the clamp nut and cable where applicable?
10. Are the connector and main identification markers appropriately identified and secured?

11. What inspection levels does vendor use in accepting the RF connectors?
12. Are the retaining ring and gasket always replaced if disassembled?
13. Are ceramic parts and metallic grooves inspected 100 per cent under adequate magnification?
14. Are adequate handling considerations given to protecting connectors? In storage? Tightening? Detorquing? Use?

25. Communication/Navigation Radio Checklist

Leo J. Boarts, Jr.
Fred Quinn
Narco Avionics

General

Electrical Design Considerations:

Have the following characteristics been adequately defined, and does the design meet the specified limits?

1. Transmitter
    - *a.* output power
    - *b.* output frequency
    - *c.* modulation limits
    - *d.* low-line performance
    - *e.* drive limits
    - *f.* duty cycle
    - *g.* switching characteristics
    - *h.* spurious responses

2. Communications Receiver
    - *a.* receiver sensitivity
    - *b.* S+N/N ratio
    - *c.* RF balance
    - *d.* image rejection
    - *e.* selectivity
    - *f.* AVC flatness
    - *g.* squelch operation
    - *h.* receiver gain
    - *i.* AGC

j. low-line oscillator performance
k. harmonics

3. Navigation Receiver
   a. HF-IF-LF image rejection
   b. ident filter rejection
   c. Nav signal accuracy
   d. VOR flag output
   e. Omni AGC—allowable course deviation for change in signal level
   f. reciprocal error limits
   g. course width limits
   h. localizer output specifications
   i. receiver gain
   j. receiver sensitivity
   k. AVC flatness
   l. low-line operation
   m. Omni left-right stability
   n. flag sensitivity
   o. to-from centering

4. General
   a. RFI and EMI considerations
   b. overload/spike/surge protection
   c. operating stability
   d. receiver input protection
   e. case detuning: Will installation cause degradation of parameters tuned out of case by necessity?

Mechanical Design and Considerations

Have the following been adequately defined?

1. Parts adequately specified—materials, finishes, tolerances?
2. Tolerances—effect of worse-case conditions. Will allowable tolerance limits require assembly fixturing to achieve desired end result?
3. Have parts been specified for flatness, deburring, etc., where required?
4. Stamping parts from proper side will eliminate need for some deburring.

Materials—Material Cost vs. Application/Use Cost

1. Do specified tolerances exceed vendor production process capability?
2. Have commercially available packaged units been reviewed to determine whether there are standard vendor items that can be used?

Process/Methods

1. Use of printed wiring where applicable to stabilize production processing, reduce costs.
2. Can design be reproduced on production scale with current manufacturing skills and capabilities?
3. Are PC boards laid out for automation?

Installation Considerations

1. Will functions be affected by changes in power source within specified limits?
2. Is unit packaged for ease of installation?
3. Is unit protected from accidental voltage reversal?
4. Are interconnecting cables "idiot proof"?
5. Does design allow for 14/28 V. conversion easily?

Environmental Considerations

1. Will unit withstand expected environmental stresses, vibration shock, temperature?
2. Will P/C boards withstand humidity without conformal coating?

Systems Compatibility

1. Has unit been married with all possible associated equipment including competition?
2. What effect will maximum variations in allowable specification tolerances have on system compatibility?

Human Factors/Product Features

1. Do channel selectors have good positive detent action?
2. Dial vs. odometer read-outs?
3. Have channel selector torque limits been defined?
4. Is panel lighting compatible with other products or competition in terms of color and intensity?

Packaging

1. Are minor adjustment controls located in such a manner as to be accessible after installation?
2. Are thermal shunts utilized to minimize the effect of heat producing components?
3. Is unit packaged to allow ease of service and maintenance?
4. Does design represent optimum simplicity?
5. Does packaging lend itself to automated subassembly testing?

Design/Marketing

1. Will design meet the established goal for quality levels, reliability, serviceability?
2. Will the design meet the required mean life?
3. Can the product be installed and serviced by existing service shops with existing equipment?
4. Does the design give the customer what he wants for the price he wants to pay?
5. Do styling and appearance meet customer's approval?

Specification Considerations

1. AVC time constant—*Comm*—very short, Receiver recovering after transmit for rapid fire talk, NAV long for Omnistability.
2. Selection of oscillator—minimize combinations possible/unuseable frequencies (fundamental and harmonics) channels, NAV to NAV, Comm to Comm, NAV to Comm, Comm to NAV.
3. Adjacent channel rejection—*Comm* currently 50 KC spacing, near future 25 KC spacing/ideal wide nose, narrow skirt. *NAV* currently 100 KC spacing, future 50 KC spacing, adjacent channel rejection critical, spillover cause erroneous navigation information.
4. Strict sensitivity control—min. and max., NAV and Comm, many dual installations one compared to other, plus competition. Unit internal coax part of match, one antenna/2 rec.
5. Automatic vs. manual, mike gain adjust, gains for microphones very inconsistent when new and deterioration with age, especially carbon types.
6. Serviceability/maintainability—most avionic sales are made by sales and service organizations.

7. Environments—severe variations and quick change, complicated by two classes of users, heavy and light (active vs. inactive plus a number of planes in winter storage) day-light/darkness.
8. Input voltage variations—generator or alternator equipped, communications needed at low RPM. Aircraft to aircraft similarly equipped voltage difference.
9. Spikes/surges/ripples— on A line, various aircraft equipment changes develop line spikes/surges and some radiation problems, such as alternator shine, etc.
10. Spurious signal rejection—flight patterns go through very strong radiated signal areas such as TV and FM stations.

11. Printed circuits vs. point-to-point—process defects much less on printed circuit boards, design performance much more constant, assembly and tool costs comparisons.
12. Transmitter/receiver spurs — multioscillators/mixers radiate on more than desired frequencies/filters/traps/critical tuning needs to minimize.
13. Critical/difficult process—competence/knowledge/ability and personnel varies—day, week, month, year. Changes, attritions, attitudes, etc. (See Checklists in Volume I.)
14. Systems, air, and ground compatibility—unit/system interconnect/interact directly, indirectly by intent and accidently. Regulations may exist on criteria/limitation.
15. Subjective criteria—brilliance of panel lighting, readability of significant information, audio response of receiver(s) and transmitter. One pilot instrument landing is similar to the notorious one arm paper hanger. NAV Receiver 10 KC attenuation and 200 cycle out of response.

16. Legal requirements—FCC and FAA, transmitter frequency tolerance, receiver local oscillator radiation limit, navigation accuracy, etc.
17. A line protection—fuse vs. slo-blow fuse vs. circuit breakers internal (part of unit), external (aircraft panel).
18. Potential shorts—usual practice, high generating devices such as power transistors are mounted externally on rear panels, sides of remote units to take advantage cooling effect of ambient, shell of transistors are not ground potential. Protection such as plastic cover may be needed.
19. Test equipment capabilities.

20. Intended vs. anticipated use.
21. Materials application.

Circuit Considerations

1. Have design objectives been established regarding minimum and maximum electrical performance throughout environmental operating range?
2. Have design objectives been established regarding physical size, weight of unit?
3. Is the NAV/COMM to be made with circuit boards or is point-to-point wiring used. (See Checklist 36 in this volume)
4. Is the NAV/COMM capable of being produced and tested as modular subassemblies? Can subassemblies be combined without special selection? (See Checklist 39.)
5. Can subassemblies be readily removed for service?
6. Are components labeled on circuit boards?
7. Are grid patterns used to lay out circuit boards?
8. Are components designed for circuit board mounting used in point-to-point wiring circuits?
9. Do component leads have to be formed or deformed to fit into a circuit board?
10. Will forming or deforming of leads break hermetic seals on semiconductors? Crack plastic packages?
11. Are Field Effect devices adequately protected from static discharges during installation? During normal handling?
12. Is heat sinking required for semiconductors? Has lead cutting reduced device power dissipation to point where prolonged operation over-stresses the device?
13. Will ceramic filters produce unwanted electrical noise/microphonics?
14. Does proper circuit operation depend on selection of matched parts? If so, do part specifications call for matched parts?
15. Are fuse time constants adequate to handle surges without opening? Over entire environmental range?
16. Are circuits adequately protected against power supply transients?
17. Are power supplies adequately decoupled between different stages/sections?

18. Are circuits/components adequately rated to handle "turn on" and "turn off" transients? (See Checklist 32.)
19. Has semiconductor leakage and its effect on circuit operation been considered?

20. Are integrated circuits operated at their maximum specified operating speed in digital frequency synthesizers? Are "race" conditions possible? (See Checklist 43.)
21. Were relay contact materials chosen on basis of type load being switched?
22. Will plastic parts withstand changes in environment without cracking, breaking, discoloring, softening, cold-flow, etc.?
23. Are mating metal parts compatible, i.e., are they close on electromotive series table?
24. Are ultrasonically welded plastic parts adequately specified regarding weld integrity?
25. Are interconnect cables properly keyed to prevent insertion of wrong plugs? Are cable connections made before electrical connections are made? (See Checklist 23.)

26. Are switches adequately specified for type load carried? Are bifurcated contacts used?
27. Are arc suppression networks used when switching inductive loads?
28. Can transmitter transistors withstand open or short circulated antennas? Withstand other VSWR conditions? (See Checklist 41.)
29. Are adequate strain reliefs provided for components on PC boards? (See Checklist 26.)
30. Are insulating materials used compatible with production process chemicals/solvents?

31. Are special tests spelled out on components specifications? Do the special tests require special test equipment/fixtures/procedures?
32. Is a special test requirement a part of procurement contract?

26. Component and Part—Terminal Board Checklist

W. B. Rossnagel

The techniques of mounting components (relays, connectors, switches, etc.) and parts (resistors, transistors, diodes, I.C.s, etc.) vary from hand and machine assembled to crimped, hand, or machine soldered. There are dozens of MIL specs available—yet sometimes the quality of imported commercial products shows better design, assembly, and processing of parts into P. C. boards.

This checklist intentionally makes heavy reference to other documents. Part of its message is just that—to alert persons to the wealth of practical design and shop level documentation available on the subject.

Here too is an example of where varying product requirements justifiably provide a wide range of interpretations as to what is necessary and what is desirable.

The Following Items Apply Particularly to Parts Mounting on Fiberglass Laminate Circuit Boards

1. Terminals on circuit boards shall be rejected if:
   - *a.* Flanges of the staked base extend beyond the periphery of the terminal area or conductor pattern.
   - *b.* Split in base of the terminal extends into the tubular shank of the terminal.
   - *c.* Two or more cracks are present in the staked side of the terminal.
   - *d.* Terminal is bent or deformed.
   - *e.* Spreading of the terminal base is not sufficient, causing the terminal to rotate with light finger pressure.
   - *f.* Soldering cannot meet the requirements of the applicable specification(s).

g. Evidence of conductor separation from the base material is present.

2. Staking of funnel eyelets shall be rejected if:
   a. Flared heads are split so that the split extends into the tubular shank.
   b. Eyelets are bent or deformed by tooling.
   c. Staking is insufficient to retain eyelet (prior to soldering) causing it to rotate with light finger pressure. There shall be no evidence of endplay.
   d. Soldering cannot meet the requirements of the applicable specification(s).
   e. Evidence of conductor separation from the base material is present.
3. Are clips used for aluminum circuit boards with transistors cases insulated from the case and board if required?
4. Is the staked side of all funnel eyelets on the part side of board?
5. Are any transistors mounted by leads alone? This is taboo in all prime equipment.

6. Are any transistor leads soldered, spliced, or extended directly to other leads (part leads or wire) without terminal support?
7. Are all requirements of MSFC-STD-154 met?
8. Are all transistors in prime equipment mounted securely to the board?
9. Is sufficient lead length available to allow for heat sink clip when soldering transistors and diodes?
10. Are you sure a single-sided board can't do the job which calls out a double-sided board?

11. Can all parts be mounted on one side eliminating the need for feed-through eyelets?
12. Are boards designed for the most simple, standard termination for relays, filters, capacitors, etc., mounted on the board?
13. On volume quantity boards, are machine insertable hardware and parts utilized wherever possible?

14. Are proper swaging tools (rolling punch, flare punch, spinning anvil, spinning tool, anvil adapter, etc.) available? Are they properly set? Are well posters displayed which illustrate correct settings and final appearance?

Module-to-Card Installation

1. Are modules installed on dummy cards per configuration sheet? Is revision correct?
2. Are modules positioned on card per configuration sheet? Is revision correct?
3. Are modules:
   a. flush
   b. without out-of-tolerance overhang
   c. damage free
   d. spacers removed
   e. flash visible at module gates?
4. Is soldering free of:
   a. birdcaging
   b. fractured joints
   c. solder splash
   d. solder beads
   e. flux?
5. Does soldering meet applicable requirements of:
   a. MSFC-158
   b. MSFC-154
   c. ABMA-PD-S-1c
   d. MSC-ASPO-S-5
   e. MSC-ASPO-S-6A
   f. QWS-10.00A
   g. AMCR 11-26
   h. AFSCM 70-5
   i. AFSCM 310-1
   j. AFSCM 375-1
   k. NPC 500-1
   l. MIL-STD-881
   m. NAVSO-INST-4130.1
   n. NAVSO P-2457
   o. 4120.3-M
   p. NHB 7121.2?
6. Do wiring connections:
   a. provide service loop
   b. stress relief?

Card-to-Rack Installation

1. Are card inserts:
   a. free of damage
   b. free of bits of wire, solder, or other foreign matter
   c. properly swaged
   d. without exposed copper
   e. bottoms and/or protective caps removed
   f. pins not touching heat sink surface?

2. Are insert cards:
    - *a.* not binding on standoffs
    - *b.* bond of card to heat sink acceptable
    - *c.* securely fastened
    - *d.* evenly seated and parallel
    - *e.* provided with card puller tool
    - *f.* secured from lifting out?

3. Unless otherwise specified, are fasteners secured with following torques:
    - *a.* #1 1.6 in-lbs torque
    - *b.* #2 2.5 in-lbs torque
    - *c.* #4 5.5 in-lbs torque
    - *d.* #6 9.5 in-lbs torque?

4. Wiring bundles. (See also Checklists 23 and 31.)
    - *a.* Spot ties per spec.
    - *b.* Is lacing card forbidden?
    - *c.* No wires allowed to be overstressed or taut.
    - *d.* No wires loose out of bundle.
    - *e.* Is wire harness neat?
    - *f.* No interlacing of wires between bundles allowed.
    - *g.* Cables protected from abrasion.
    - *h.* Cable supports/hangers provided where necessary.

5. Cable and Harnesses. (See Checklist 23.)
6. Coaxial Cable Connectors. (See Checklist 24.)

27. Electrical Assemblies and Component Checklist

Courtesy of Rocketdyne
a Division of
North American Rockwell

This checklist has been taken directly from a Rocketdyne Division of North American Rockwell Corporation Design Manual. It represents only a small portion of the checklists used by this large aerospace operation. Rocketdyne's checklists show an important consideration for the sequence in which the points are asked. This checklist consists of two sections labeled "Assemblies," and "Components."

Assemblies

1. Can access to components be achieved without damaging cables or soldered wire connections?
2. Are terminal strips readily accessible?
3. Have provisions been made for extra terminals? Added components?
4. Will these be required?
5. Are all terminal strips and component boards properly marked in a permanent manner?
6. Are safety factor allowances adequate on all components for all environmental conditions? (See "Work Statement and Environmental Checklist" in Vol. VII.)
7. Has integrity of all solder joints been evaluated under combined vibration, acceleration, and thermal conditions?
8. Are solder methods, fluxes, etc., properly controlled to ensure quality?
9. Are all wires and cables properly protected against abrasion or injury? Where they pass through partitions or components? In close proximity to structures?

10. Has the proper assembly sequence been established?

11. Are there any special tools or fixtures required?
12. Can they be eliminated?
13. If not, will they be defined in assembly, test, and maintenance specifications? Will they be brought to GSE's attention early?
14. Are any components in close proximity to extreme heat or cold?
15. Is shielding required?

16. Is insulation required? Has type, availability, and procurement specification been established?
17. Are heaters required?
18. Has reliability of the type to be used been established?
19. Are power requirements for heaters provided?
20. Will thermostatic controls be used?

21. Is control reliability adequate?
22. Have the results of heater failure on the components been considered? On the system?
23. Have wire sizes and other electrically conducting parts been calculated for ability to carry:
    a. nominal loads
    b. peak loads
    c. Under most adverse environment? What safety factor exists?
24. Have proper sealing methods been used where necessary? (See also "Seal Checklist" in Vol. IV.)
25. Can any components be installed backward?

26. Can provisions be made to prevent misinstallation?
27. Can connectors be attached to the wrong components or receptacles? Has connector clocking been coordinated?
28. Can installation error be avoided by different size connectors?
29. Have all identifications and markings been added where required?
30. Are there any unnecessary joints or splices?

31. Are cables looped properly to drip away from connectors?
32. Does the position of connectors and other components preclude the entrance of water?
33. Can system reliability be improved by the addition of redundant circuits? Components?

34. Have you considered the assembly for space environments of:
    a. vacuum
    b. radiation
    c. explosive decompression?
35. Have provisions for overload protection been made on all circuits?

36. Will the assembly cause radio interference? Be affected by it?
37. Are there any special assembly, maintenance, handling, or storage requirements?
38. Are these adequately documented?
39. Have all components in this assembly been individually evaluated? (See Checklist 26.)
40. Do all wire and cable lengths allow for various installation tolerances?

41. Have provisions been made to adequately secure all mechanical elements, fasteners, threaded, or bolted assemblies? Are torque values specified?
42. Have corrosion protection considerations been made on all mechanical elements?
43. Will the functional test specification include adequate checkout of all wiring to preclude miswiring, poor solder joints, cracked or damaged insulation, broken wires, etc.?
44. Have all leads been evaluated on the installation:
    a. to avoid close proximity to extreme temperature
    b. to avoid crushing, pinching, or rubbing damage due to actuation of other components (e.g., gimballing, etc.)?
45. If explosive devices are used, has adequate consideration been given to wiring, accessibility, sealing, protection, etc.?

Components

1. Does this component possess contact which may be subject to vibration failure, wear, corrosion (e.g., relays, switches, etc.)?
2. Have solid state vs. mechanical elements been evaluated for this component?
3. Has the most reliable from a performance standpoint been selected?
4. Did you consider operation under the worst environmental con-

dition? (See "Work Statement and Environmental Checklist" in Volume VII.)

5. Did you consider service life? Storage life?

6. Have you considered performance at minimum voltage? High potential? Flashover?
7. Will it emit radio waves?
8. Will it be affected by radio interference?
9. Will its operation spark an explosion in an explosive atmosphere?
10. Is it in itself explosively inert?

11. Can it withstand explosive shock?
12. Will it be affected by gamma radiation? If so, what is its maximum allowable exposure time to probable maximum density?
13. Can this component be easily installed?
14. Are there special assembly, maintenance, handling, or storage requirements? Will these be controlled in procurement specifications, drawings, or process specifications?
15. Are any nonstandard parts or components being used?

16. Does your development test program provide for establishing the reliability and integrity of this component in this application?
17. Can parts or the component be installed backward, upside down, wired wrong, etc.?
18. If this is injurious to operation, can it be avoided?
19. Has component connector been coordinated as to type and clocking for installation?
20. Will the hard vacuum of outer space affect the material or operation of this component? Will rate of vacuum application have an effect?

21. Have you considered the effects of increased friction? Trapped gases?
22. Is it possible to damage this component in installation by dripping solder, bent pins, dropping, etc.?
23. Are explosive squibs, initiators, or igniters used in this design?
24. Is the milliampere fire current characteristic considered sensitive (25 to 1,000)? If so, is this necessary?
25. Have you considered the effects of environment on the initiator? (See "Environmental Checklist" in Vol. VII.)

26. Effects of long storage?
27. Can RF energy fire the initiator? Static charge? Stray currents? Heat?
28. Can these be avoided?
29. Where cables or harness must permit flexing during installation or servicing, have the insulation, wire length, joints, etc. been evaluated for ability to take such flexure without damage or failure?
30. Was this considered at worst possible environments?

31. Are electrical heaters, where used, compatible with GSE requirements?
32. Have the effects of solenoid inductive kickback been considered?
33. Has adequate protection been provided from overloads or short circuits?

28. Electrical Design Checklist

Courtesy of the Boeing Company
Seattle, Washington

This is a Boeing checklist, excerpted from Boeing Document D2-9476-1, "Design Reliability Manual-Electrical, Mechanical and Fluid Flow." It is an excellent checklist on electrical design considerations because it is specific and concise. Like most of the others in this manual, the items carry a message as well as raise points to be considered.

The author has personally witnessed Boeing personnel utilizing checklists such as this one in conducting critical design reviews on both Lunar Orbiter and Minuteman III. These checklists are good, but in the hands of skilled engineers who are not too good to use it, it can represent a formidable review and as negotiating one tool.

Circuits

1. Are system circuit tolerances adequate?
2. Are circuits designed to prevent shorts caused by high voltage and overloads?
3. Are the power source, signal, grounding, and shielding adequate for the circuit design?
4. Are printed circuits qualified to the full range of expected environmental conditions?
5. Are power circuits designed so that components such as motors will not operate needlessly whenever external power is applied?
6. Are circuit disabling devices (interlocks) provided on equipment where maintenance must be performed on subassemblies connected with but remote from a central control panel?
7. Do temperature sensing, monitoring, controlling, or warning

circuitry provide fail-safe features in closed cooling system to prevent equipment overheating and failure in the event of cooling or circuit failure?

8. Are time delay circuits provided to control high voltage during gyro and filament warm-up, and sensing components to excessive voltage variations?
9. Is the design of a time delay circuit such that its protective function is not nullified by installation of a manual bypass switch?
10. Have the effects of variation in power supply (voltage, frequency, pressure) on the system been investigated?
11. Do test notes specify that whenever verification tests are conducted on the temperature sensing, monitoring, controlling, or warning circuits, such tests shall cover not less than the entire circuit loop, and never a section thereof?
12. Are systems (such as missile, erection and/or firing systems) electrically deactivated during functional checkout to prevent inadvertent operation?
13. Are protective interlocks in the system so designed that they will not be bypassed when relay simulator cards are inserted?

Dials

1. Are dial pointers so installed that they cannot become loose on shafts?

Lights

1. Are light assemblies of the continuous operation type with spare bulbs available in permanent holders?
2. Is operating voltage low to minimize failure?
3. Are lamp assemblies shock mounted where possible to avoid failure from vibration?
4. Will circuit breakers or fuses carry full load when lights are turned to full brightness?
5. Has the possibility of color selection for lights to be used for different purposes and functions been considered?

Motors

1. Has it been verified by realistic test that electric motor failure will not be attributed to:

- *a.* overheating due to motor starting torque to overcome load
- *b.* bearing failure
- *c.* excessive brush wear (use brushless components where possible)?

Potentiometers

1. Has adequate consideration been given to the following common causes of variable pot failure:
   - *a.* open and shorted windings
   - *b.* erratic outputs (rough spots on windings; conductive particles, generated internally, lodging between wiper arm and resistive element)
   - *c.* wiper arm binds or hangs up
   - *d.* internal solder contamination generated by excessive, externally applied heat?

Relays and Switches

1. Has the desirability of using present electromechanical relay devices in any airborne control and switching circuits been considered?
2. Are relays not hermetically sealed protected from freezing during altitude cycling?
3. Will contact resist chattering due to vibration?
4. Has the buildup of high resistance on contacts caused by metal breakdown due to arcing and contamination on contacts been investigated?
5. Are adjustments for relay contact gaps (power circuits) protected from misadjustment due to vibration and improper maintenance?
6. Has the use of contactless "can" type relays, as used in printed circuits, been considered for low-voltage circuits?
7. Has an arc-suppression network been installed across the contacts to absorb the magnetic surge and reduce contact failure?
8. Vibrational forces can induce rotational movement of the coil bobbing and result in lead wire breakage. Has adequate consideration been given to prevent considerable movements of relay parts?

9. Have the contacts been sealed or isolated from contaminating vapors and organic materials?
10. If the design is of a nature that contact enclosure is not possible, have lubricants and organic materials in surrounding atmosphere been tested for possible contact contaminating effects?
11. Is the vendor contributing, in handling, shipping, or during assembly, to the contact contamination problem?
12. Has an adequate evaluation test program been conducted utilizing simulated operational limits to assure long contact life?
13. Has a positive requirement been established for a hermetic seal of a switch or relay?
14. Does the switch configuration lend itself easily to positive actuating and release action?
15. Could the shaft-end bearing surfaces of an electrically actuated rotary switch, subjected to wear, contribute to possible misalignment and a need for excessive drive force?
16. Have all internal insulating materials of a hermetically sealed unit been analyzed to determine if contamination is possible during operating and use?
17. In a sealed unit, is it possible for internal wiring to interfere with any movable parts?
18. Could transient-suppressor components be mounted externally to a sealed unit?
19. If transient-suppression components are used within a sealed unit, will the internal operating temperature have any degrading effect on the suppression component?
20. Are the terminals of a sealed unit adequately identified to prevent the possibility of misapplication of coil polarity voltage?

Part Selection Considerations

1. Have proven components, circuits, and assembly designs been considered where new designs and materials do not represent a major design improvement or when available performance and test data fail to show that a higher reliability will result from the new component?
2. Have vendor-improved processes and variations of the *known* design been considered?

3. Are all components and parts adequately qualified to environmental, electrical, and mechanical parameters within which the installed components are expected to operate?
4. Are adequate tolerances allowed in such component applications as component temperature rise and subassembly ambient temperature rise, cooling any electrical overloads, frequency deviations, transients, and changing component values with usage?
5. Has vendor quality control practice been considered in vendor selection?
6. Has the application of preferred circuits such as described in NAV-AER 16-1-519 been considered to increase circuit stability and equipment MTBF?
7. Have standardization of parts and uniformity of hardware sizes been established?
8. Can assemblies containing movable parts function with the minimum of lubrication?
    a. Are lubricants compatible with air frame temperature cycling and environment changes?

Parts Integration

1. Is adequate clearance allowed between externally mounted (electronic, electrical, etc.) assembly connectors?
2. Are control devices on the outside of subassemblies well defined and separated from those required for maintenance or adjustment?
3. Are internal circuit diagrams and technical ratings clearly and permanently marked on all potted or canned components (transformers, relays, etc.)?
4. Has the most suitable type of marking application been used?
5. Are terminal boards, chassis, controls, panels, connectors, etc., clearly and permanently marked with circuit symbol and purpose of function of each switch or control position to avoid faulty operation resulting in damage?
6. Do structural members or other components allow direct flow of cooling air to heat generating components (tubes, resistors, etc.) within air cooled subassemblies?
7. Are fuses used in subassemblies easily accessible so that damage will not result during replacement?

8. Does each fuse have a spare provided in a permanent holder?
9. Has the need for special tools been minimized?
10. Have supply problems resulting from the use of nonstandard parts been considered?

11. Are all instrument dials, indexes, and markings finished with luminous paint?

Parts Integration—Construction

1. Is design of subassembly shape and weight such that installation in limited access and crowded areas will not cause difficult maintenance conditions or damage?
2. Are handles provided on all apparatus where size, weight, or shape will affect ease of installation, transport, or maintenance? (Most electrical apparatus weighing more than fifteen pounds require handles or obvious gripping surfaces so that damage due to improper handling will not occur.)

Parts Integration—Placement

1. Are components and subassemblies mounted in areas subject to possible adverse environmental conditions sealed, protected by sealed covers, or mounted in such a way that water, fuel, and dirt cannot enter unit?
2. Does adequate clearance exist between subassemblies and structural supports so that damage cannot occur to electrical and cooling connections when vibration of the subassembly takes place?
3. Has it been verified that subassembly or structural damage cannot be caused by the following installation conditions?
    a. installing equipment in areas subject to damage by storage of equipment and manuals during ground maintenance
    b. installing equipment in areas subject to damage sustained by misuse of equipment (example: using equipment as a step)?
4. Are installations of components with rotating parts near heat source shielded against heat to prevent evaporation of the rotating part lubricants?
5. Have vibration mounts been protected from deterioration due to exposure to ozone or hydraulic fluid?

6. Has it been verified that improper system responses will not occur by the installation of magnetic-sensitive assemblies in areas subject to coupling to other assembly components or systems?
7. Is each subassembly accessible and removable from its installation without interference with other structures and subassemblies?
8. Is the illumination adequate to perform maintenance without the aid of hand-held flashlights, etc.?

Component Hold-down

1. Will all vibration mountings support the weight of the subassembly without bottoming?
2. Are screws flush with the rear surface of mount so they cannot protrude into cable or other damageable parts?
3. Do all fastener assemblies contain adequate locking components and/or possess practical but effective torquing requirements?
4. Are all subassembly fasteners (nuts, bolts, etc.) easily accessible to maintenance personnel so that malfunction or damage will not result from faulty installation?
5. Will all fasteners or nut plates support the weight of the unit under stress conditions?
6. Are fasteners of low magnetic permeability?

Interconnection—Wiring

1. Has it been verified that the wiring, coax cable, and wire bundle problems will not be a cause for nuisance rejections, improper system response, or subassembly or system failures?
2. Is the protection of wires and cables passing over and through partitions or through lightening holes adequate to prevent insulation wear and breakage due to wires rubbing on metal surfaces?
3. Have as many connectors and terminal strips as possible been eliminated from cable and wire runs to reduce the possibility of intermittent or open circuits or other changes to circuit characteristics (impedance)?
4. Will slack in cables or wires
    a. allow for structural flexing or temperature expansion
    b. eliminate stretching and pulling of wires or cables when disconnecting and connecting service subassemblies?

5. Are wires or cables rooted near heat sources heat resistant or shielded?

6. Has decomposition of insulating materials been considered?
7. Will all clamps used in supporting cabling withstand the vibration environment in which they are to be used and not damage wire insulation?
8. Is stranded instead of solid conductor used in all subassemblies where wire flexibility is required?
9. Are specifications, which provide minimum handling controls essential for reliability, called out on cable-connector assembly and installation drawings?
10. Should warning notes on drawings point out highly vulnerable electronic assemblies?

Interconnection—Connectors

1. Is sealing of connectors adequate in areas of environment-induced contamination such as dirt, moisture, fuel, etc.?
2. Will shielding and soldering at connectors prevent open and shorted circuits?
3. Will wire and cable run design prevent broken wires at plugs?
4. Are keyways and connector guides provided to minimize bent and broken pins and broken inserts caused by blind mating of connectors with loose inserts?
5. Are tightening requirements specified on drawing call-outs?

6. Are drip loops or potted connectors used where moisture is likely to enter plugs? Drip loops also assure adequate slack.
7. Has the mechanical design provided for the stresses produced by high cable weight?
8. Have the extra torque stresses generated during coupling and uncoupling been considered?
9. Has the elimination of very small contact pins been considered?
10. Has the use of several smaller cables been considered as opposed to a single larger cable?

11. Has missile structure strength in umbilical plug installation areas been evaluated, and all stress requirements considered?

12. Are specifications, which provide minimum handling controls essential for reliability, called out on cable connector assembly and installation drawings?
13. Should warning notes on drawings point out highly vulnerable electronic assemblies?

Interconnection—Bonding

1. Are bonding requirements adequate to give a good conductive path for electrical assemblies?
2. Are mating metal surfaces, which are required to be clean, properly identified on drawings?

Alignment—Adjustments

1. When adjustments are necessary, is easy access provided?
2. Do all adjustments have provision for positive locking?
3. Are adjustment values, such as wave forms, voltages, current, or power, called out at test points?

Power Supply Cables

1. Are insulation or cables of proper design and material to withstand damage from water, abrasives, footsteps, vehicles, and other abuse when lying on the ground or floor?
2. Are racks provided on equipment for coiling and hanging power cables when not in use?
3. Are power cables so located that personnel will not tend to grasp them for support?
4. Is cable sufficiently flexible to perform required function?
5. Does the design incorporate permanently attached connector dust caps such as chain-attached or spring-loaded types for use on cable ends or exterior panel receptacles?
6. Does the design include removable dust caps for all rack and panel connectors or other connectors not provided with permanently installed caps?
7. If rack and panel connectors are used:
   a. Do tabulations of maximum tolerance buildup in dimensions of both drawer and rack assure that:
      (1) connector alignment and blind mating capability will remain unimpaired even with maximum tolerance accumulation

(2) when drawers are fully installed, the mated connectors will remain free-floating and will not be required to bear any portion of drawer weight?

b. Does a check of connector retaining plate cutout dimensions and tolerance, including corner radii, assure that the full required ranges of free-float have been allowed for and will remain unobstructed even with maximum tolerance buildup?

8. Are process specifications, which provide minimum handling controls essential for reliability, called out on cable connector assembly and installation drawings?
9. Should warning notes on drawings point out highly vulnerable electronic assemblies?

29. Electrical/Electronic Reliability Checklist

W. B. Rossnagel

Many checklists have been written covering the reliability aspects of electrical and electronic devices. This checklist is a compilation of the best inputs from more than a dozen checklists used in both the commercial and military electronics industry.

In most cases, the inputs to this item were utilized in design reviews and proposals to customers. Following the completion of this checklist in 1968, it was distributed to half a hundred companies where the majority also included it in the sales, training, and manufacturing/inspection planning work.

1. Has the effect of reliability tradeoffs been determined?
2. Have the proper Hi-Rel, MIL, or commercial grade parts been specified?
3. What is the previous MTBF or MCBF? Is it acceptable?
4. What is the MTBF or MCBF design objective? Has either been calculated?
5. Is life testing or accelerated aging of parts required?

6. What are the failure modes? Is a failure mode and effects analysis required?
7. What are the effects of failure? Is this a flight safety item?
8. Have the effects of failure been minimized in circuitry?
9. Will component or system failure cause other items to fail?
10. Have fail-safe techniques been considered?

11. Would any redundancy features help reliability?
12. Has complexity been kept to a minimum?
13. By what rules have components been derated in power, voltage, current?
14. Have key reliability margins (deviations) been determined?
15. Does the design meet the reliability objective?

16. Have derated failure rates been determined?
17. Are confidence levels applied to a numerical reliability requirement?
18. Can components be easily replaced or overhauled without removal or damage to other items?
19. Can equipment be inspected and/or tested without dismantling?
20. Has necessity for adjustment been minimized?

21. Is ventilation a problem? What is actual temperature of hottest spot?
22. Where forced air cooling is used, are suitable dust filters incorporated? (See "Thermal Checklist" in Vol. III.)
23. Is rating of relays switching inductive or incandescent loads adequate?
24. Have relays been properly derated and protected? Is a no-miss test specified on relay drawings for Hi-Rel applications? (See "Relay Checklist" in Vol. III.)
25. Has proper attention been applied to transistor circuitry leakage problems? At high temperature too?

26. What EMI/RFI spec is applicable? (See Checklist 32.)
27. Does design reduce EMI and RFI interference to a minimum?
28. Is overload, spike, short circuit, and surge protection provided?
29. Have components and wiring, which might be affected by magnetic fields, been shielded from iron core components?
30. Has the use of friction or pressure contacts been reviewed?

31. Do signal input wires need to be shielded and/or twisted? Are they properly separated?
32. Does spec require special isolation or grounding of parts?
33. Do production work standards require heat sinking of components when soldering?
34. Have easily accessible test points been provided?
35. Have interlocks been provided where necessary?

36. Is circuit stability sufficiently within tolerance over entire operating range? Are there any oscillation problems?
37. Have all functional inputs been considered?
38. Have realistic performance tolerances been specified?
39. What is the maximum internal temperature rise? Is a means of detecting improper operation incorporated?

40. Are silicon transistors specified over germanium except for very special applications? Does plant "burn-in" transistors or does vendor?

41. Have trimpots been properly selected for CRV? What is minimum adjustment change required (in m.v.)?
42. Where printed circuits are used, are they qualified to the full range of expected environmental conditions? Does PC design meet or exceed requirements of MSFC-154? (See Checklist 36.)
43. Has use of time-delay high-voltage or regulated heater circuits been considered?
44. Are power transistors properly heat sinked?
45. Is adequate clearance provided between P.C. boards, cables, parts, connectors, etc.? (See Checklist 23.)

46. Have sufficient service loops been allowed in wiring? Is a Hi-Rel, soldering spec specified?
47. Is all wiring routed clear of heat sources or protected from heat?
48. Are contact terminals of type with good history?
49. Is commutator transfer and rise time adequate?
50. Has SBT (select by test) been kept to a minimum?

51. Is the circuit design as simple as possible?
52. If the design is similar to previous designs:
    a. Was the design of good quality
    b. Was the design trouble free
    c. Was it easily built, adjusted, inspected, tested?
53. Does the bill of material (B/M) agree with the schematic?
54. Does the schematic agree with the wiring diagram?
55. If relays are necessary, are contact ratings sufficient for the anticipated loads? Is spike suppression required?

56. Is input and output impedance matched with other subsystem components?
57. Are TBD (to be determined) part callouts kept to a minimum?
58. Is there a mix of Hi-Rel/MIL spec/commercial parts? Are Hi-Rel parts called for when MIL-spec parts could be used consistent with component application and criticality (and vice versa)?
59. Are there any areas where a part is used beyond 50 per cent of its rated capacities? Are parts sufficiently derated for high-temperature operation?

60. What preconditioning is specified for critical part types?

61. Are standard modules used with standard pin matrices?
62. Is part packaging density reasonable? Is module easily manufactured? Easily tested? Easily inspected?
63. Is the method of attachment of the modules to a printed circuit board clearly defined? If bonding agents are utilized, is sufficient definition given?
64. Is a clear encapsulating material used wherever possible to facilitate troubleshooting and rework?
65. Are modules constructed as complete test circuits to facilitate testing and reduce complexity of test equipment?

66. Are special handling fixtures required for unpotted module handling? Will plugging of unpotted modules into test sockets cause stresses which might break welds, etc.?
67. Can the module/component be easily adjusted/tuned? Are adjustments required after potting? After final assembly?
68. Is it easy to replace parts or modules if they fail?
69. Can troubleshooting be easily performed?
70. Can module/component be easily assembled?

71. Is there a minimum of blind assembly conditions which often cause creased wiring, shorts to chassis, etc.? Are thick bundles of wires avoided?
72. Is the method of insulating power from case adequate and foolproof?
73. Is signal ground common to power ground?
74. Has consideration been given to the effect of potting on:
    a. distributed and stray capacitance in RF circuits
    b. shrinkage and subsequent stresses on parts
    c. other circuit effects?
75. Are there sufficient adjustment features available?

76. Can adjustments (tuning) controls be locked after final trim-up?
77. Can individual sections of redundant components be singularly adjusted and tested?
78. Are sufficient test points available for in-process tests?
79. Are critical final test points brought out to external connectors so that a final "in-case" check can be made?
80. Can a failure mode(s) of a minor circuit abort the entire system function unnecessarily?

81. Is the design foolproof as contrasted with trying to foolproof the user?
82. Do any unsafe conditions exist?
83. Does the component performance specification contain adequate definition of all required parameters? Does the ATP? Does the QTP?
84. Are all performance requirements listed with the required tolerance limits for each? Can you test all parameters?
85. Are drawings available? Are they complete? Why not?

86. Do you have all the documentation to perform *your* job?
87. Is jitter significant in nanosecond pulse measurements?
88. Have you established a rule not to break the test setup on a gigacycle component until the failure can be repeated?
89. Have basic troubleshooting hints been provided to others who can benefit from them?
90. Is phase rotation indicated where applicable?

30. Electrical/Mechanical GSE/AGE/TSE Checklist

H. J. Williams
Kearfott Division of Singer

The problems associated with the testing of large subsystems and systems at one or more field sites reached the pinnacle with the Apollo Program. With the Minuteman, Polaris, and each of the aircraft developed in the last two decades has come a continual growth in the complexity and intricacies of cable and harnessing, semiautomatic and digital testing, as well as magnetic and punched tape inputs to multibay fixed or portable test racks. Some of these included hydraulic, pneumatic, vacuum, air conditioning, cooling, and other inputs in addition to the complex electrical test sequences.

As a result of ten years work in design and test reviews, resolving manufacturing problems, participating in customer buyoffs, and solving the many field interface problems, a checklist was evolved on this subject. When used in conjunction with other checklists, which were contained within it, it represents the single, most effective review document for all levels of management in all aspects of ground support equipment, aerospace ground equipment, and test support equipment.

General

1. Are all affected support items completely compatible with the design requirements of the aircraft, vehicle, or equipment?
2. Have human factors inputs been incorporated into the test and support facilities for this item? (See "Human Factors Checklist" in Vol. V.)
3. Is the affected design adequate to prevent (or greatly minimize) accidental damage to the aircraft, vehicle, or equipment even under potential misuse? (See "Safety Checklist" in Vol. V.)
4. Will a potential malfunction in any way damage the aircraft, or equipment?

5. Will the aircraft, vehicle, or equipment in any way become contaminated during normal (or abnormal) use?

6. Are all drawings and specifications involving use of the test and support facilities up to date? Are they adequate? Are they being used?
7. Are all cables, hoses, etc., available and in proper condition?
8. Are all drop shipped items included in inventory?
9. If unit requires special handling, are complete instructions contained in the maintenance manuals?
10. Can the unit be readily installed and connected to the aircraft, vehicle, or equipment? Are instructions available?

11. Are factory adjustments such that they do not require readjustment when units are replaced in a system or when parts are replaced in the unit in the field?
12. Can every fault (degrading or catastrophic) which can possibly occur in the unit be detected by the use of the proposed test equipment and standard test procedures?
13. Does hoisting equipment contain adequate provisions for preventing interference of cables with stationary parts?
14. Are bottom sheaves of hoists equipped with close fitting guards to prevent the rope cable from dislodging?
15. Are warnings placed on each item of hoisting equipment indicating the maximum allowable load in pounds or tons?

16. Are cranes, hoists, and other lifting equipment fitted with a braking apparatus capable of effectively braking a load in compliance with established governmental regulations or other proof load requirements?
17. Will the reeving of wire rope used with hoists preclude excessive stresses on level wind mechanisms?
18. Are all fasteners at critical locations safety wired or provided with cotter keys or self-locking nuts?
19. If equipment is mobile, have stresses which will be encountered during transportation been considered in addition to those which are encountered during normal operations?
20. Can methods other than S hooks be used on hoists, slings, and other lifting equipment? (These hooks are inherently unsafe and may result in lack of control unless the hook is subject to tensile load at all times.)

21. Is MIL-STD-202 referenced or used as guide for all applicable tests?
22. Does the equipment have an elapsed time indicator to assure periodic maintenance at proper intervals?
23. Are safety factors of two or greater applied to limiting characteristics of all parts?
24. Is each control, indicator, and test point functionally identified, and are they grouped logically to simplify operation? (See "Human Factors Checklist" in Vol. V.)

Electrical

Design

1. Are limiting resistors used in test point circuitry, i.e., is there protection against component failure in the event that a test point is grounded or an adjustment resistor is turned to its extremities?
2. Are transistors, capacitors, vacuum tubes, and diodes that are fixed to panels, circuit boards, and chassis locations free of mechanical resonances in the 9 to 500 Hz. range?
3. Are fail-safe circuits designed so that they can only be reset by authorized personnel?
4. Has equipment been designed to allow for expansion? For containing key spares?
5. Have sufficient troubleshooting features been designed into the equipment to allow for detailed analysis of the hardware under test?

6. Has the equipment been designed so that it can be calibrated?
7. Has anyone checked with the calibration laboratory to insure that adequate standards of the proper value and tolerance to perform the required calibration are available?
8. Are calibration standards traceable to the National Bureau of Standards? Do they meet all of MIL-C-45662A? (See checklists in Vol. V.)
9. Has a 10:1 tolerance ratio been maintained between the test equipment and the parameters under test, wherever possible?
10. Is the equipment electrically safe? Are all voltage points which could constitute a personal hazard protected with either a placard or an interlock? (See "Safety Checklist" in Vol. V.)

11. Can the test equipment operator willfully or unknowingly damage the unit under test by improper operation of the test equipment? (Is it idiotproof—Murphy's Law?)
12. Is the portable or movable equipment designed so that it can be moved without tipping (extended bases, low c.g., etc.)?
13. Are all pulse generation or shunt circuits calibrated?
14. Does a detailed calibration instruction exist for all test equipment?
15. Is there a system in place to prevent the use of uncalibrated test equipment?

16. Is "out of calibration" equipment removed from the test area immediately? Or labeled as "out of calibration"?
17. Are "as is" calibration runs made whenever equipment is turned in for calibration?
18. Is the test equipment user notified whenever an item turned in for recalibration is found out of spec?
19. Does test equipment measure all jumpers and monitor loops?
20. In addition to good grounding practices, are static and magnetic field sources and noise sensitive circuits shielded from one another to prevent interaction?

21. Are sources of conducted and radiated electromagnetic radiation adequately filtered and shielded to prevent external interference? (See Checklist 32.)
22. Are all trim (calibration) controls limited in adjustment range to the maximum expected variation of the adjusted characteristic?
23. Has the effect of all possible combinations of inherent measurement errors been considered, including the loading effects of the test equipment?
24. Are the connectors which are to mate with the unit under test the most durable possible, and are they properly mounted in the test fixture?
25. Is the connector keying for all cables designed to prevent improper interconnection of equipment?

26. Are all cables supported so as to prevent stress concentration at the connectors, and are external cables protected from traffic damage?
27. Is the maximum use of common electronic modules and subassemblies considered?

Wiring (See other checklists in this volume.)

1. Are all harnesses entering and leaving a panel or chassis firmly clamped as close as possible to the point of entry or departure?
2. Is maximum use made of existing holes for clamps in a panel or chassis?
3. Are harness bundles wrapped or tied and clamped as required? Or is procedure per the following schedule:
   a. harness diameter 1/8"-1/2"... maximum clamp spacing 6"
   b. harness diameter 1/2"-1"... maximum clamp spacing 8"
   c. harness diameter over 1" ... maximum clamp spacing 10"?
4. Are clamps generally installed on harnesses if the following conditions exist:
   a. a possibility of flexing of small breakout wires that could cause the wire to fatigue and break
   b. a flexing of a fragile terminal that could cause the terminal to fatigue and break?
5. Are photographs taken of internal and external features of completed rack equipment?

Mechanical (See checklists in Vols. III and IV.)

1. Are approved limit switches provided on all hoists and lifting devices?
2. Are storage space and tie-downs (or compression packing) provided in the equipment to store accessories and cables supplied with the equipment?
3. Are sufficient handles provided, and are they centered over (if on top), or above (if on the sides) the center of gravity on equipment to be carried?
4. Is the center of gravity low, and does it stay within the "footprint" of rack-mounted equipment even if all drawers are extended?
5. Has door height or hatch size been considered in determining maximum envelope dimensions of the equipment?
6. Are all structural screws and bolts locked with lock washers or lock nuts?
7. Are all drawers in rack supported at the rear or sides with slides or braces?

8. Are all front-panel-mounted controls and indicators protected from possible damage during bench maintenance even if the equipment is placed face down?
9. Are the wheels or casters on roll-around equipment large enough —6" diameter minimum?
10. Is the equipment supplied with suitable work surfaces for data taking and unit under test where required?
11. If cooling air is provided to the unit under test, does it operate interlock switches which sense cooling failure by temperature or flow rate?

Hydraulic (See checklists in Vol. IV.)

Pneumatic (See checklists in Vol. IV.)

31. Electronic Part Selection Checklist

W. B. Rossnagel

Also see the "Microelectronic Process/Quality Checklist" (44), and the "Electrical/Mechanical GSE/AGE/TSE Checklist" (30).

First, (*a*) Has vendor approved part drawing? (*b*) Is a specification control drawing required? (*c*) Is a source control drawing required?

Resistors

1. What is operating ambient temperature? Maximum ambient temperature?
2. What power dissipation is estimated?
3. Is the resistor properly derated? In power *and* voltage? By what factor? Are reliability rated resistors required?
4. What tolerance limit is required for satisfactory circuit operation? Has a worse-case analysis been performed?
5. Is resistor tolerance affected by temperature, aging, electrical stress, etc.?

6. Can any potentiometers be replaced by fixed resistors?
7. Have the following factors been applied to resistors which dissipate more than 1/10 rated power (if necessary)?
   a. clamp resistors to a heat sink
   b. use short leads
   c. mount resistors in free-flowing air
   d. mount power resistors in groups vertically
   e. resistors over 5 inches should be mounted vertically
   f. prevent power resistors from radiating heat to sensitive parts.
8. Have resistors for space hardware been adequately derated? In a vacuum environment, is conduction the only available way to take away heat?

9. Is temperature coefficient adequate? Is it linear?
10. Are high-voltage resistors kept clean and free of fingerprints, and other oven contaminants?
11. Are resistors mounted per MSFC-154?
12. Is aging or burn-in required? Is lot traceability required?

Capacitors

1. What is operating ambient temperature?
2. What working voltage is expected?
3. Is the capacitor subject to surge voltages (particularly start-up or shutdown surges), operating voltages? Will ripple voltage combine with d.c. voltage to exceed rating?
4. Is the capacitor properly derated? In voltage, frequency, and temperature?
5. What tolerance limit is required for satisfactory circuit operation? Has tolerance buildup been considered? Is it OK for plus tolerance to exceed negative tolerance? Or vice versa? What precautions are taken to assure reverse voltage is not applied to polarized capacitors?
6. Are tantalum capacitors bypassed for high frequencies (above 100kc)?
7. Have temperature-compensating or low-temperature-coefficient capacitors (mica or ceramic) been used whenever high stability is required?
8. Are radiation shields necessary?
9. Are heat sensitive capacitors isolated from hot parts?
10. Are NP capacitors utilized, or two capacitors back-to-back (polarized)? Are reliability rated capacitors required?
11. Capacitors are heat sensitive in the following order of decreasing sensitivity:
    - *a.* electrolytic
    - *b.* mylar
    - *c.* tantalytic (foil and slug)
    - *d.* paper tubular (wax impregnated)
    - *e.* phenolic molded paper
    - *f.* oil filled, sealed, metal clad
    - *g.* ceramic

h. teflon
i. glass and mica

12.. Have high-dielectric ceramic capacitors been restricted by bypass usage?
13. Are physically large capacitors mounted in accordance with the applicable specification?
14. Is p.f. critical in any tantalytic application?
15. Are capacitors mounted per MSFC-154?
16. Will hydraulic or bending leads cause mechanical damage?

Diodes

1. Does specification for the diode type define the required characteristics? Is steepness of "knee" of curve defined?
2. What is the operating ambient temperature range for each diode?
3. What is the power dissipation within the diode? What is maximum rated power dissipation?
4. What is the breakdown voltage? What is maximum forward current?
5. How much reverse current can be tolerated?

6. Is the inrush usage current exceeded for capacitor input?
7. What is the rated peak inverse voltage?
8. Is reverse recovery time for the diode critical?
9. How much reverse current will flow at the operating ambient temperature?
10. What Zener voltage reference is expected and required? Is the "knee" of low voltage Zener's specified?

11. Does the circuit perform satisfactorily with randomly selected diodes, and at their upper or lower limits?
12. Can diodes be either barrier coated or uncoated?
13. Are rectifier fins mounted vertically, and are the fins of an optimum size?
14. Is a low thermal resistance to heat sink provided? How?
15. How are diodes to be packaged after receiving inspection?

16. Plastic, epoxy, and silicone encapsulated diodes caused a number of critical failures in 1968. Have they been carefully selected if used?
17. Are glass diodes coated with black paint to eliminate light sensitivity? If so, how is interior visually inspected?

18. Can a Unitrod nail-head diode (no open tail line mode) be used to replace a ribbon or whisker type at no additional cost premium?
19. Are X-rays of diode function and whiskers desirable? Necessary?

Transistors

1. Does the specification of the transistor type define the required characteristics (hfe, Icbo, Vcbo, fT, etc.)? Also Vebo, Vebx, others? When?
2. What is the operating ambient temperature range?
3. Has a low-thermal resistance path to heat sink been provided for power transistors?
4. What is the maximum rated power dissipation, and what is the maximum power dissipation expected in this application?
5. Does the operation of the transistor approach any absolute rating under any usual variation of supply voltage or load?

6. How much voltage derating is present?
7. Are spikes or transistors present in application requiring additional voltage derating?
8. What is the maximum collector voltage, and how does it compare to the present application?
9. What is the maximum rated collector current?
10. What deviation in beta is tolerable, and over what temperature range? Is beta selection or matching required?

11. How much deviation in beta is expected from aging? Radiation?
12. Will the circuits perform satisfactorily with transistors randomly selected and at their upper or lower limits?
13. Have high-frequency transistors been tested at high frequency and load or merely at d.c. conditions? If only at d.c. potentials, is worst-case analysis made?
14. What deviation in power gain is tolerable and what deviation is expected from tolerance build up?
15. Is the noise level tolerable at the operating ambient temperature? How does noise vary over operating temperature range?

16. How much leakage current is expected at the operating ambient temperature? At maximum operating temperature?
17. Are *germanium* devices excluded from any ambient temperature exceeding 16° F?

Vacuum Tubes

1. Does the specification of the tube define the required characteristics? For Class *A*, *B*, or *C* operation?
2. Does the operation of the tube approach any absolute rating under any specified variation of supply voltage or load?
3. What is the operating ambient temperature?
4. What electrode ratings are critical in circuit application?
5. Is the heater voltage within rating? What voltage variations are expected?

6. Should unshielded tubes be spaced at least 1 1/2 diameters apart?
7. Is the heater-to-cathode voltage within rating in circuit application?
8. Are the plate and screen grid dissipations properly derated?
9. Were maximum grid-resistance ratings observed?
10. Has Gm variation been considered? With temperature? With age?

11. What variation in input and/or output interelectrode capacity can be tolerated?
12. Does circuit operation depend on a tube parameter not controllable by the circuit designer? If so, feedback may be required?
13. What are the maximum rated and the maximum expected bulb temperatures?
14. Have adequate cooling measures been taken to prevent damage to P.C. boards (when used), or components? (See Checklists 23 and 24.)
15. Have standard tube shields been used? What effect do they have on bulb temperature and tube life? Are the shields necessary or just for appearance uniformity?

16. Is suppression required for any special oscillations, such as Barkhausen or ionic bombardment?
17. Does filament circuit require automatic plate and screen voltage cutoff in event of opening or failure?
18. Is a delay desirable on application of plate and screen voltages?

Relays and Switches (Including reed relays and reed switches)

1. Has the use of relays and switches been kept to a minimum?

2. How many actuations per hour, per mission, or per lifetime are expected?
3. What percent of rated resistive current does each contact carry?
4. Is relay closing or opening time critical? Is "bounce" critical?
5. What are the required pull-in and drop-out voltages?

6. What is the manufacturer's tolerance for initial coil resistance?
7. Does relay switch inductive loads? If so, is it properly derated? Is suppression provided for back EMF on contact opening? Is it realized suppression increases response time?
8. Has arc and spike suppression been considered for circuit?
9. Has the possibility of dry-circuit operation been considered?
10. Do relays require a contact resistance of less than 50 milliohms? Is low contact resistance obtained by excessive tradeoff of current rating?

11. Is the limit on switch contact resistance practical? Can it double after 100,000 cycles?
12. Is contact resistance measured "dry circuit" or under actual load?
13. Have MIL-R-6061 and MIL-R-5757D been used to specify performance, environmental test, etc., requirements?
14. Do all switches and relays incorporate a "wiping" action?
15. Have all precious metal contacts on open-type relays been properly cleaned with acid free cleaners? Are dust/moisture covers required?

16. When was the last time an X-ray was made of a representative sample of relays to discover loose parts or debris after vibration, shock, or acceleration?
17. Have proper call-outs been made for mercury relay applications?
18. Are the relays free from parts that could cause teflon/nylon whiskers?
19. Are relay sockets only used where necessary? Are they labeled?
20. Are relay test sockets properly mated? Are they specified for the relay to be used?

21. Are relays with internal connected diodes adequately marked with "caution notes"?
22. If relays switch only one side of line, is it only the "hot" side?
23. Have effects of transient suppression devices on response time been considered?

24. It is not advisable to use a lamp across a relay coil as an "operate" indication in many circuits. Has this been checked?
25. Are relay contacts designed for the frequencies switched?
26. Are latching relays spaced approximately 1/2 inch from each other? Why?
27. Is careful control held on relay housing dimensions since they constitute substantial factor in vibration resistance of relay?
28. Because of the difference in closing and opening times, paralleling of relays to share load is not advisable. Is it done?
29. Are hermetically sealed relays required?
30. On reed relays and reed switches:
    a. If contact voltage exceeds twelve volts, has an inductor or capacitor been used in series with switch to reduce stray capacitance which can create high current on contact?
    b. Is a limiting or biasing resistor used to limit current on cold state resistance?
    c. When a switch is used in a "carry only" or with loads over one amp, is a two ms. min. time lapse allowed between energization and current passage? Is a .5 ms. min. allowed between current break-off and contact break?
    d. Is an interlock used where feasible to prevent inadvertent contact opening during high-current carry?
    e. With R-C suppression, oscillation can occur on coil deenergization. Has this been checked to assure reclosure cannot occur?

Transformers, Chokes, and Coils (See Checklist 40.)

1. Are clean, low-resistance thermal paths provided to heat sinks?
2. Have thermal paths been cleaned of impregnators and paints?
3. What is the operating ambient temperature?
4. What is the maximum allowable surface temperature? Is it measured by the resistance method, or by a specifically located thermocouple, or by a paste-on temperature indicating device?
5. What deviation of Q can be tolerated? What Q deviation is expected from tolerance buildup and temperature change?
6. What is the maximum current-carrying capacity? What is the maximum current expected? Is phase shift critical?

7. Is it realized that tape-wound core transformers, saturable reactors, toroids, etc., can usually generate several hundred millivolts during vibration without any applied voltage?
8. Was the operating frequency band of the circuit compared to the resonant frequency of the choke or coil?
9. Has a requirement for shielding been established?
10. Has a computer analysis been utilized in determining design parameters?
11. Do valuable indicators have a means for locking the adjustment device?

Electromechanical Devices

Note: See checklist in Vol. III for motors, accelerators, gyros, and stable platforms, servomechanisms, and solenoids.

1. What consideration has been given to variations of the motor speed-torque characteristics due to temperature, altitude, changing load, changing input voltages, etc.
2. Have the adverse effects on brushes at high altitudes been considered?
3. Can the power supply tolerate an increased or decreased electrical load caused by variation in motor characteristics?
4. Is stall or lock rotor torque specified? (See "Motor Checklist" in Vol. III.)
5. Have the force-movement characteristics of solenoids been considered?
6. Is utilization of current trace made in studying plunger movement?
7. Have resolvers been checked for accuracy and phase shift at elevated temperatures?
8. Are meter windows adequately sealed to prevent moisture formation?
9. Are end stops provided for all limited movement equipment?
10. Are lubricants, where used, adequate under vacuum conditions at ranging temperatures?

Connectors (See also detailed Checklists 23 and 24.)

1. Has the use of cable connectors and plugs been kept to an absolute minimum? Is more than one wire allowed to any pin?

2. Does the number of active pins per connector conform to the recommended limit? Is there a minimum of 2 or 10 per cent spare pins?
3. Has the isolation of circuits (power vs. signal) at the connector been considered?
4. Has the power loss to be dissipated from each pin, under conditions of maximum acceptable contact resistance, been considered?
5. Will the pin current capacity be exceeded under any conditions?

6. Has the need for termination of shields been considered?
7. Have a uniform policy and written procedures been established for potting?
8. Are connectors potted in a manner where no voids are in the terminal area?
9. Has the use of redundant parallel pins been considered to assure minimum contact resistance in critical circuits?
10. Are high-potential gradients possible between adjacent pins?

11. Is an immersion test necessary?
12. What provisions are made to prevent and discover "bent pins"?
13. Are all connector sockets of the hooded "antiprobe" style?
14. Are pins and sockets free-floating where required to aid engagement?
15. Are connectors with built-in EMI filters desirable?

16. Are connector shells long enough to prevent handling damage to pins through mishandling?
17. Are any connectors having tuning-fork-type contact used?
18. Is wire O.D. compatible with sealing?
19. Do solder and crimp cup have viewing holes?

Printed Circuit Boards (See Checklist 30.)

Mounting Parts on P.C. Boards and/or Cards (See Checklist 36.)

1. Has placement of parts on printed circuit boards been considered from cross talk point of view?
2. Does a heat dissipation problem exist on printed circuit boards?
3. Is keying employed to prevent interchanging printed circuit boards?
4. Are transistors, diodes, or capacitors properly polarized on printed circuit boards?

32. EMI/RFI Electrical Interference Checklist

Albert R. Kall
Technical Director
Ark Electronics Corporation
Willow Grove, Pennsylvania

When the technology is as much as an art as a science, as is the case with electromagnetic interference and radio frequency interference, a checklist can be even more effective. This checklist has a three year history of compilation. It has also undergone more rewrites than any other in this volume.

Probably no other aspect of modern electronics is treated more "hit or miss" or belatedly than the interference technology. Too often the tendency has been to try and luck out with a quick fix, but for every instance this happened there are five more cases in which additional filtering has to be crammed into insufficient space with the resultant costly redesign/retest.

The government has recently released three triservice standards which replace thirteen existing electromagnetic compatibility documents. This change was long overdue.

Specification Requirements

1. Which provisions of the specifications apply? Are they followed? (See references at end of checklist.)
2. Is the test equipment required to conform to all contractual specifications, including EMI/RFI?
3. In using automatic test equipment per MIL-STD-461, is scanning speed proper? In using manual methods, is the scan rate not less than three frequencies per octave?
4. Are any special tool requirements specified on drawing (if applicable)?

Circuitry Design

1. If possible, is the equipment free of pulse circuitry?
2. Has unnecessary fast logic been used in the design?
3. Are pulse rise and fall rates under 25 volts per use, if possible?
4. Are pulse amplitudes less than 10 volts, if possible?
5. Are large pulse generators located as close as possible to their intended use area?
6. Is the reference level of the generating source traceable?
7. Are "clock pulse" circuits shielded, filtered, and decoupled? What controls are on transient and audio susceptibility and generation?
8. Are overdriven cascaded low-grain amplifiers used? If so, examine carefully for possibility of overdriving which could result in the generation of undesirable harmonics.
9. Are low-level analog circuits avoided wherever possible?
10. Are diodes and transistors free of electrical interference-generating characteristics? Are power switching relays separated from logic boards?
11. Are high- and low-powered stages separated by compartments?
12. Does the test power supply have the same internal impedance characteristics as the supply with which the equipment in the system has to operate?
13. Are blower motors of the a.c. noncommutating type, if possible?
14. Is d.c. rotating equipment designed to minimize commutation effects?
15. Is the "wave-guide-below-cutoff" type of filtering incorporated where necessary and applicable?

Oscillator

1. Is oscillator power kept to a minimum?
2. Are oscillators shielded from other stages and from antennas?
3. Do parasitic oscillations exist? Is suppression necessary?
4. Is the oscillator heater decoupled from $B+$ supply sources?
5. Is the oscillator relatively immune to loading variations or high VSWRs?

Mechanical Design/Construction

1. Has electrostatic and electromagnetic interface been isolated and/or shielded separately? By separate techniques?

2. Are sharp projections and bends avoided in high-voltage circuits? Is any corona present?
3. Are the magnetic fields associated with indicators adequately isolated?
4. If an RF seal is to be effected, then it should be pressure tight. Is it?
5. Are openings (louvers, ventilation, access, etc.) shielded to prevent case leakage?

6. Has the chassis or frame been grounded?
7. Is it necessary to bypass shock mounts with grounding straps?
8. Has the insulated protective finish been removed where a metal-to-metal contact is required?
9. Are access door grounds of the metal-textile or finger-strip type? Are they electrically grounded?
10. Do all covers have RF gaskets? Are filters incorporated within components?

11. Are cover-mounting screws spaced the correct distance (less than half the smallest wave length to be suppressed)?
12. Are mechanical components protected from strong fields of RF energy?
13. Is the use of conductive finishes for circuit continuity avoided?
14. In the construction of the chassis and other metal parts of the assembly, are metals used which are widely separated in the electromotive force series of the elements? Avoid such use wherever possible; otherwise avoid incidental contact between such different metals.
15. Have MU metal shields been provided for servomotors, synchros, resolvers, etc., where necessary? Is MU metal of proper heat treat condition?

16. Are variable impedances, such as potentiometers, tranducers, dial lamps, etc., protected against high noise or vibration?
17. Are there any potting voids around high-voltage points?
18. Are iridite or bare metal finishes used on cabinet joint or other mating surfaces? Are "sticky finger" self-adhesive beryllium copper contacts provided on openings to reduce EMI/RFI leakage?
19. Is any silver plating/flash for shielding properly protected from damage during handling and transportation? Are all silver-plated parts stored and shipped in sulfur-free boxes, cartons, etc.?

Decoupling/Isolation

1. Are all signal leads separated from power leads?
2. Is undesired signal transfer reduced by interstage decoupling networks and link- or parallel-tuned circuits?
3. Are pulse networks and transformers isolated? Is micrologic circuitry (and system) designed with additional EMI/RFI considerations which are not critical in high-level logic? Are multiplexers, decoders, etc., properly selected for filters? (See Checklist 35.)
4. What measured power supply transients will be created by relay or switch operation?
5. Are indicator control and power leads decoupled?

6. Does the design include the use of high-power sharp cut-off devices, such as thyratrons and SCRs? If so, have precautions been made to isolate these elements from the rest of the assembly by shielding, filtering leads, etc.?
7. Are a.c. heaters wired with twisted and/or isolated leads?
8. Are pulse network leads decoupled as short as possible?

Suppression/Filtering

1. Is *too much* filtering used?
2. Does any filter have a resonance or a frequency which might affect the equipment performance?
3. Are the RFI filters mounted in the correct direction, i.e. are they mounted specifically to keep RFI from getting in the box, or from getting out?
4. Has transient suppression been used on the relays?

Wiring and Connectors

1. Do connectors require "carry through" RF shielding? (See Checklist 24.)
2. Is ground discontinuity avoided at connectors? How?
3. Are power and signal leads generally separated? What consideration was given to common path impedance and bus noise cross talk?
4. Is signal ground to chassis ground kept short as possible? Is shielding grounded at proper wave lengths?
5. When is wrapped solid shielding required on cables in lieu of conventional expandable tubular shielding? When are individual or twisted pairs to be shielded?

6. When is a “drain lead” required to be incorporated in a bundle of shielded wires?
7. Are all tests for EMI/RFI run with proper lengths of interconnecting cables?
8. Are all power and control leads, subject to frequent switching transients, properly shielded and routed separately from other wires?
9. Has an attempt been made to duplicate vehicle harnesses in the test cable setup?
10. Are shield terminations as short as possible?

11. For a shielded multiconductor cable, has proper design attention been given to the interface problem between the cable shield, shield termination, and the connector?
12. Are transient carrying wires routed away from low-level signal wires within the equipment?
13. Are there any ground loops or common node ground returns in the test setup?
14. Is EMI emitted from interface between cable and back of connectors? Is new “Sunbank” design used?
15. Is pulse energy always fed to succeeding stages via coaxial leads where possible? Is a check made of wave form distortion caused by coaxial cable connectors and capacitance?

Tests

1. Is the equipment being tested in all the same operating modes as it will operate in the system?
2. Is the equipment being tested as start-up and turnoff as well as under continuous operation?
3. Are ambient interference baseline runs to be made before and after the equipment measurements?
4. Is the facility grounding system noise free?
5. Does the screen room provide the necessary shielding? How many db. down?

6. Does the test setup have a static ground?
7. Shouldn’t complete ground plane breadboard tests be run?
8. Have tests for electrical noise been performed? Were scope photos made?
9. Has test equipment for electrical noise been performed? Were scope photos made?

10. Is this test rule followed? "The test specimen should be the ground node point, no other nodes should exist."

REFERENCES

"Electromagnetic Compatibility Prediction Studies, Tech. Report RADC-TR-66-560".

"Electromagnetic Reliability and Compatibility." TOD 70-12.

"Interface Notebook" (AD 484585), Defense Documentation Center.

"MIL-STD-449C, Radio Frequency Spectrum Characteristics, Measurement of."

"*MIL-STD-461A, Electromagnetic Interference Characteristics, Requirements for Equipment."

"*MIL-STD-462, Electromagnetic Interference Characteristics, Measurement of."

"MIL-STD-463, Electromagnetic Interference Characteristics, Definitions and System of Units."

"MIL-STD-469, Radar Engineering Design Requirements, Electromagnetic Compatibility."

*These triservice documents supersede the following.

MIL-STDS	MIL-SPECS	Other
"MIL-STD-826"	"MIL-I-6181"	"NFEC-SPEC-50Y"
	"MIL-S-10379"	"AF/BSD Exhibit 62-87"
	"MIL-S-12348"	
	"MIL-S-16910"	
	"MIL-I-17623"	
	"MIL-I-26600"	
	"MIL-I-43121"	
	"MIL-I-55301" (E)	

33. Flexible Cable and Circuitry Checklist

Courtesy of
Methode Electronics Inc.

This checklist is taken directly from the manufacturer's advertisement. It represents the absolute *minimum format* in checklist information, but it is an effective way to attract and reach the reader. Very few persons would read the ad and not note several advantages in the product which they had not previously been aware of—advantages which affected their selection.

Toward this end, checklists open many avenues of improving technical advertising copy in the endless task of trying to reach the customer. This checklist has purposely not been expanded, as have others in this manual, in order to illustrate the effectiveness of brevity. (See also Checklists 23 and 24.)

1. Are the following mechanical characteristics specified?
    a. conductor thickness
    b. conductor width
    c. center-to-center distance
    d. edge-to-edge distance
    e. insulating material
2. Does the design use a standard pattern where possible?
3. Have the capacities, if substantial, been derated sufficiently to prevent overheating as a result of stacking or multiple usage of optimum currents?
4. Are cables supported at adequate intervals and locations?
5. Are cables shielded from hot components?

6. Are cables routed to avoid interference with component installation?
7. Is there sufficient extra length to provide for easy plug-in and removal of connectors?
8. Are cables protected from sharp corners and projections?

9. Are the overall size and the individual conductor sizes and spacings compatible with available connector types?
10. Have splicing operations been planned for use of eyelets or rivets, welding, soldering, brazing, or other methods?

11. Have cost-versus-reliability concepts been studied to permit decision as to use of termination with:
    a. printed circuit terminator cards
    b. circuit boards
    c. connector plugs with encapsulating boots
    d. specially sealed environmental connectors?
12. Have the following environmental and operating conditions been evaluated in the application?
    a. current carrying capacity and voltage drop
    b. overload rating
    c. dielectric strength
    d. capacitance
    e. impedance
    f. shielding
    g. vibration and shock requirements
    h. flex-life and bending radius
    i. temperature extremes, high and low
    j. flammability
    k. temperature cycling
    l. high altitude
    m. humidity cycling
    n. fungus resistance
    o. chemical and solvent resistance

Note: Have the applicable portions of the following specifications been noted?
a. MSFC-Spec-219A Connectors Flat Conductor, Flexible Electrical Cable
b. MSFC-Spec-220B Cable, Flat Conductor, Flexible, Electrical Copper

34. Laser Checklist

John W. Gillespie, Vice President
Jodon Engineering Associates, Inc.
Ann Arbor, Michigan

The following lists operating parameters, physical characteristics, and warranty considerations which should be considered and/or specified when selecting a helium-neon gas laser. While these considerations are specifically applicable to helium-neon lasers, they are, in general, also applicable to other ion lasers such as argon, krypton, and helium-cadmium.

Parameter or Characteristic	Specification Units
1. Type plasma	Helium-neon
2. Type excitation	DC or RF
3. Power output	Milliwatts, minimum, at specified wavelength and mode, i.e. 15. Omw, at 6328 Angstroms and TEM_{OO}
4. Transverse Mode	Uniphase (TEM_{OO}) or multiphase
5. Longitudinal or axial mode spacing, C/2L	MHz; typically 197 MHz for 15. Omw laser
6. Wavelength	Angstroms or microns, i.e. 6328 Angstroms (visible red), or 1.15 microns (infrared)
7. Beam diameter at 1/e2 points	Millimeters; typically 1.5-1.2 mm for 15.Omw laser
8. Beam divergence, full angle	Milliradians, typically 0.7-0.9 mm for 15.0 mw laser
9. Beam polarization	Linear, or linear to better than 1 part in 10^3 typically
10. Polarization angle	$\pm$ angle to vertical; typically with in $\pm$ 5° of vertical

Parameter or Characteristic	Specification Units
11. Resonator configuration	Mirror design; long radius hemispherical
12. Resonator cavity length	cm; typically 75-80 cm for 15.0 mw laser
13. Plasma tube type	*a.* Brewster windows or internal mirrors *b.* coxial center fed or end fed *c.* Hot or cold cathode
14. Amplitude stability, long term (after warm-up)	± percent; typically 3-5 percent after 1/2 hour warm-up
15. Beam amplitude noise (1-100KHz)	2-3 percent RMS
16. Beam amplitude ripple (120Hz)	0.5-1 percent RMS
17. Power supply (laser exciter)	*a.* Solid State design *b.* Regulated *c.* Input voltage, current and frequency; typically 117 or 234 VAC, 1.4-1.8 amperes, 50/60Hz, for 15 mw laser *d.* Output voltage; typically 6.3 VDC filament; and 3000 VDC at 25 ma HV for 15 mw laser
18. Environmental limits	*a.* Operating temperature, ° C; typically 10-40° C *b.* Storage temperature, ° C; typically 40° C to + 80° C *c.* Humidity, percent RH; typically 90 percent RH *d.* Altitude, feet; typically sea level to 10,000 feet
19. Mechanical Dimensions	
a. Laser head	LxWxH in inches and weight; 32x4x4 inches, 20 pounds for 15 mw laser
b. Power supply	LxWxH in inches and weight; typically 10x5x7 inches, 14 pounds for a 15 mw laser

Parameter or Characteristic	Specification Units
c. Power cable length (laser head to exciter)	Feet; typically 6 feet
d. a.c. cable length	Feet; typically 8 feet
e. Beam Height above table	Inches; typically 3 1/4 inches, adjustable ± 0.5 inches using mounting feet
20. Options	*a.* Alternate mirror coatings for 1.15 and 3.39 micron infrared outputs *b.* Large bore plasma tubes for increased 1.15-3.39 micron outputs.
21. Warranty	Mechanical, electronic, and optical parts, including plasma tube; degree of warranty and duration; typically, warranted against all mechanical, electronic, and optical assemblies, to be free of defects in workmanship and materials for one (1) full year following delivery. Warranty not normally including glass breakage. Performance warranty generally defined as laser output will be equal to or greater than 1/e X specified minimum power output throughout the warranty period.
22. Price	Verify price of laser and exciter combination, with FOB point defined Consider cost of, out of warranty laser servicing, particularly plasma tube refilling or replacement, and the cost of ca-

Parameter or Characteristic	Specification Units
	vity cleaning and alignment Consider cost of laser accessories and optional equipment
23. Delivery	Number of days or weeks, after receipt of order; typically, 20-30 days ARO
24. Service provisions	Factory or service representatives, location of service centers, service period required for refurbishment and/or plasma tube replacement
25. Method of shipment and FOB point	Define means of shipment, shipping weight, number of separate shipping containers, deposit for reuseable containers, and FOB shipping point

Note: *e* — Natural Logarithm Base (2.718)

35. Multicoder/Multiplexer Checklist

Gene Lieberman
Operations Manager
and
Earl Newborne
QC Manager
Fifth Dimension Inc.
Princeton, New Jersey

Here is an example of a checklist which is utilized by a firm's engineers in design reviews, in customer technical interchange meetings, and as part of technical literature supplied by sales to customers.

It contains that necessary detail and scope which caused it to be selected as one of the electrical/electronic checklists which so aptly illustrates the "Checklist Concept." With it, a supplier can make it easier for both experts and quasiexperts to understand the various performance requirements imposed on the multicoder or multiplexers.

Design and Specifications

1. Have the following performance characteristics been considered and specified?
   - *a.* Back Current—What is the back current in nonamperes, both differential and single ended?
   - *b.* Common Mode Rejection—What is the common mode rejection, in decibels, a.c. and d.c.?
   - *c.* Cross talk—What is the interchannel cross talk for both normal signal ranges and over voltage?
   - *d.* Offset between Channels—What is the channel-to-channel offset in percent of full scale referred to the output?

e. Linearity—What is the linearity in percent of the full scale best straight line?

f. Output Noise—What is the noise in percent of full scale referred to the output?

g. Frame Rate—What is the frame rate and tolerance in percent of nominal?

h. Input Voltage Range—What is the input full scale voltage range?

i. Maximum Power Input—What abnormal overvoltage, undervoltage, reverse polarity, transients arise, for example, from transducer failures?

j. Gain(s)—What gain is required? Tolerance? Should gain be adjustable?

k. Duty Cycle—What is maximum standby and full on-duty cycle?

l. Imput Impedance—What is the input impedance, differential, and single needed?

m. Output Impedance—What output impedance is necessary? Specify output load limits.

n. Pedestal and/or Output Code—What output code is needed? For example, PAM, PCM, PDM?

o. Sampling Rate Stability—(See frame rate.)

p. Environments—What operating and nonoperating environments are present, such as temperature, altitude, humidity, shock, vibration, acceleration, salt spray, fungus, etc. (See checklist in Vol. V.)

q. EMI/RFI—What requirements are to be met? (See Checklist 32 in this volume.)

2. Have the specifications for all the performance characteristics been clearly defined? With tolerances or limits?
3. Has the design considered the specified environment, i.e. fungus resistant materials? Sealed for humidity?
4. Does the end item application and/or specification and/or contract require the use of Hi-Rel, Mil spec or commercial parts? (See Checklist 31.)
5. Has the use of dissimilar metals been avoided? (See checklists in Volume VI.)
6. Are weight and dimensional requirements critical?

7. Has protection against accidental reversal of power line polarity been provided?
8. Has protection against signal overvoltage to the limits of the specification been provided?
9. Has protection of all output circuits been provided in case of short circuit?
10. What effect do input, output, or power line cable lengths have on performance characteristics?

11. Are these cable lengths critical?
12. Are programmable gain or frame rate characteristics, if required, independently stable from each other?
13. Are input and output impedances compatible with overall systems requirements?
14. Is any "select by test" of electronic parts really necessary?
15. Are adjustments provided to allow user to "tailor" parameters to specification in his end equipment? Will they be stable under age and environments?

16. Has any analysis been made to assure all components are within their rated tolerances under all environmental and input power conditions?
17. Can any combination of input voltage, environment, and customer adjustments cause damage to the unit?
18. Is the necessity for, and frequency of, periodic testing or operation by the customers specified? In electromechanical multicoders, distribution of lubricant in bearings is a consideration under extended storage. (See "Mechanical/Electromechanical" Checklists in Vol. III.)
19. Is the customer speaking the same language when dealing with your specifications?
20. Does output data require conditioning for use with transmitters (low pass filters)?

21. Have the main frame and subframe channel sampling relationships been selected to minimize logic circuit complexity?
22. Does the input require signal conditioning?
23. Is the input gating device normally open when power is removed from the multiplexer?
24. Is the frequency response of the signal conditioner equal to the requirements of the sampling rate?
25. Can failure of a single input gate affect the operation of other gates? If so, how many?

26. In cases of bipoler inputs, does the signal conditioner have sufficient off-set stability for the specifications required?
27. With PCM, is the data to be presented as unitpoler binary, octal, BCD, Excess 3, Gray Code? With bipolar one's, two's complement?
28. Is data (PCM) output serial or parallel?
29. Does the PCM format require odd or even parity bits?

Test and Inspection

1. Have measurement limitations for these characteristics been considered, i.e. noise output limited to 10 KC bandwidth and sampling rate stability over 10 period average?
2. Is the method of measurement of all specified parameters standard? If not, is a separate, detailed test procedure supplied?
3. Has a clear differentiation been made as to which of the performance characteristics are design goals and which are actual test requirements? Whether they apply at room temperature or at hot and cold environments?
4. Have conflicting tolerance requirements between different tests been avoided, i.e. to specify 0.1 per cent linearity with 0.5 per cent noise is conflicting since the noise will mask out any attempt to accurately measure linearity.
5. Have similar output requirements been clearly defined to avoid confusion?
6. Has the correct EMI/RFI spec been listed? Are test conditions defined? Cable lengths, shielding? (See Checklist 32.)
7. What workmanship standards are to be used in manufacture? Inspection? (See checklists in Vol. V.)
8. What special process standards are utilized in manufacturing? What special finish requirements are applicable?
9. In case of an open channel, does the output stabilize at some predetermined level? What level? (normally negative limiting)
10. Is any special test sequence required for environmental testing?
11. Is multicoder operation required during any phases of environmental conditions?
12. Has the length of time required for "temperature stabilization" of equipment been defined?
13. Will unit operate at extremes of temperature and voltage input specifications?
14. Are requirements specified to the easiest acceptable limits?

15. Are any adjustments necessary when installing or replacing the multiplexer?

16. Is provision of an auxiliary synchronization pulse required or necessary for measurements?
17. Will unit operate with full power applied instantaneously to it?
18. Will unit begin operating with power raised slowly from zero to full? At what voltage?
19. Docs thc output format meet the specification in regard to form (PAM, PCM, PDM) and coding (IRIG, Special)?
20. Is the sync gap amplitude specified, and are there channels designated for calibration voltage outputs?

21. With gated PAM format, is the data dynamic range specified? The pedestal plus off-time reference levels specified?
22. For PDM, are the minimum and maximum pulse widths specified in relation to the input signal levels? Also, are the minimum and maximum pulse widths specified for positive and negative overvoltage inputs?
23. For PCM format, is the output wave form specified? NRZ, NRL, Manchester?

36. Printed Circuit Board Checklist

Robert L. Lang, Jr.
Purchased Material Supervisor
Narco Avionics

Printed circuits now comprise the basis of mounting parts and forming circuitry in virtually all electronic products. In addition, they usually include precious metal contacts for mating with card holders or connectors. They are unique in that many phases of technology are deeply involved in their design and construction, such as metallurgy, chemistry, photography, mechanical engineering, plastics, and electrical engineering as well as almost every process and quality activity in the manufacturing plant.

This checklist is typical of the detailed information/question type checklist which can be utilized by virtually all persons specifying, designing, buying, manufacturing, or testing this product.

Separate mention is made of the last section on multilayer or microprinted circuit boards which are now reaching a point of high demand, particularly in the aerospace and defense industries.

Selection of Laminate Materials

1. Is there a note on the drawing which states, for example:
    a. etch printed wiring per artwork
    b. material—glass epoxy NEMA G-10, 1/16 thick, 2 oz. copper both sides
2. Does the laminate specification for moisture absorption show sufficiently low percentage? Does it agree with customer specifications? Have specs on page 1 of MIL-STD-275B been utilized?
3. Does the laminate resist distortion from heat? From punching?
4. Is the laminate fungus resistant or at least not a fungus nutrient?
5. Will the laminate withstand the standard etching and plating solutions used in conjunction with printed circuitry?

6. If the finished part specification demands it, will the laminate withstand vibration, flexing, shock, and humidity without delaminating? Is the accept/reject criteria clear?
7. Is the plastic laminate material marked in such a manner that it will be identifiable after etching and dip soldering processes?
8. Is the grade of copper used of sufficient quality and thickness to prevent occurrence of pinholes and surface contaminations?
9. Is the minimum thickness of laminate being used compatible with the needed mechanical rigidity? What is tolerance on thickness?
10. Is the dielectric constant and power factor of the laminate acceptable at the frequencies that will be used in the circuit?
11. Is the copper bonded to the laminate thick enough to allow sufficient carrying capacity for a reasonable width of etched conductor?
12. Will the bond between the copper and the base laminate withstand heat and humidity environments without separating? Is it tested?
13. Does the specification for the copper laminate call out a peel strength test agreeable to the finished P.C. board customer?

Circuit Design

1. Does drawing clearly indicate:
   a. number of holes
   b. hole size and tolerances
   c. hole location, preferably on a grid base
   d. overall dimensions and tolerances
   e. number of layers
   f. who supplies artwork
   g. thickness of dielectric between layers
   h. dielectric material
   i. minimum spacing
   j. conductor width (minimum)
   k. conductor thickness?
2. If tolerances on photomaster are reduced, will this result in additional inspection time and costs? Can tolerances be held?
3. Has the master layout been made on a low coefficient of expansion material such as Kronar or glass?

4. In the selection of line widths and the spacing between lines, do these dimensions agree with the military or customer specifications?
5. Are the spacing and the voltages between lines compatible with the altitude at which the board will be expected to function?

6. Are the power leads placed at the board periphery?
7. Are the circuits distributed on the board in such a manner as to minimize capacitive coupling and cross talk? (See Checklist 33.)
8. Are holes, unless otherwise dimensioned, located on voids in the pads; sized as indicated, and located within .007 inch radius of true position?
9. Does the conductive pattern extend at least .020 inch from outside edge of hole?
10. Is the conductive pattern coated with a minimum of .0003 thick solderable composition consisting of at least 60 per cent tin content?

11. Is line width .030 minimum, and spacing between conductive areas at least .020 after solder coating, unless otherwise specified? Is there a minimum of .030 copper around each hole?
12. Are all "plated through holes" drilled, and have 100 per cent plated hole walls of not less than .002 thick fine grain ductile copper plating under the coating in Note 10?
13. Are all hole sizes specified as finished sizes after plating?
14. Is alignment between conductive patterns on opposite sides of a board within .010 inches in any direction?
15. Will any conductors be damaged by tool used to swage eyelet terminations?

16. Are all exterior edges and sharp corners chamfered at least .010 inch?
17. Have each of the P.C. boards been identified by the manufacturer with a batch or lot number that will identify date of manufacture?
18. Are you sure a single sided P.C. board cannot do the job? Is a two-sided board really necessary?
19. If a two-sided board is used, is the layout between the two sides utilized efficiently to allow grouping of circuit functions?

20. Is the initial layout (before photography) at least four times larger than that of the final circuit so errors may be minimized?

21. Do the designed holes for component leads have a clearance dimension of at least .007 inches?
22. Has the layout been accomplished on a 0.1 inch grid system or to customer specification?
23. Have free flowing lines and rounded corners been utilized?
24. Has the layout been made "point to point" or as short wiring as possible?
25. Have large copper areas, where there is no circuit, been checkered or crosshatched to reduce the possibility of blistering during soldering?

26. In general, have high-frequency and high-impedance circuits on etched wiring boards been avoided?
27. Has consideration been given to the use of finger holes, where applicable, to facilitate safe removal of the boards for maintenance?
28. Are applicable requirements of the following specs met?
    a. MSFC-154B-Design, Documentation, and Fabrication of Printed Wiring and Printed Circuit Boards
    b. MIL-STD-275B—Printed Wiring for Electronic Equipment
    c. MIL-STD-429 Printed Wiring and Circuit Terms and Definitions
    d. Institute of Printed Circuitry Inc. Standards
    e. Code: #1001-(Dwg #XWS6337)-BUWEPS—Gen. Spec for Multilayer Printed Wiring Boards

    f. MIL-T-10727—Tin Plating, Electrodeposited and Hot-Dipped
    g. MIL-P-13949 Plastic Sheet
    h. MIL-C-14550 Copper Plating, Electrodeposited
    i. MIL-G-45204 Gold Plating
    j. MIL-I- 46058 Insulating Compound Electrical (for P.C.)

Part Mounting (See also Checklist 26.)

1. Are any parts mounted so close to edge of board as to interfere with jigs or wave soldering holding fixtures?

2. Are the component mounting holes spaced in such a manner as to allow interchangeability of different manufacturer's parts at the assembly line?
3. Are all components larger than a one watt resistor secured to the board by other than their electrical leads?
4. Where several components meet at a common mounting point, have separate mounting holes been provided?
5. Are components which overlay other terminals insulated?
6. Is the layout such that there is at least 1/16 inch space between any conductor and the outside edge of the finished board or support structure?
7. Have holes for component leads been spaced at least 1-1/2 times the laminate thickness apart and from outside edges and cutouts?
8. Has space been provided for support points at intervals of not more than 4 inches for boards 3/32 inch thick and less?
9. Has the use of eyelets been avoided if the board is to be subject to vibration?

Shop Processing (See also Checklist 26.)

1. Are all burrs, lint, and traces of machining compounds removed?
2. Is the shop elongating the holes beyond detailed dimension to facilitate assembly?
3. Is the shop countersinking, chamfering, or drilling holes at an oblique angle to board?
4. Are part leads crimped over on underside of board?
5. If a pneumatic cutting tool is used to clip off straight through uncrimped part leads, are you sure it is not damaging soldering or conductors on the board or shocking or stressing critical parts?
6. Is a temperature log kept on solder pot or wave solder bath?
7. Are timers controlling solder time closely followed in manual dipping operation?
8. Is there any evidence of too little heat and/or joints incompletely wetted?
9. Plating features shall be rejected if:
    a. plating is not smooth, continuous, and uniform in color
    b. plating contains a defect that will impair solderability
    c. base metal under plating is exposed.
10. Is the flux cleaning operation adequate, and carefully controlled?

11. Are completed P.C. boards packaged in sealed plastic bags?
12. Do boards show any evidence of blistering, delamination, measling, or crazing after solder dipping in 60-40 composition in accordance with applicable NEMA, IPC, or government standards?
13. Is finished board ink or paint marked with latest effective revision letter at time of manufacture? Marking material shall not be affected when cleaning with Kester #AP20 Flux Remover.
14. Is the quality of all plating and solderability in accordance with the "preferred" category of the Institute of Printed Circuitry?
15. Are finished boards individually packaged to provide protection during shipping and for a minimum of six months storage?

16. Can boards be machine wave soldered and still pass the contour solder requirements of MIL-STD-275B? Will wave soldering cause rejections for excess solder as established in this spec unless otherwise defined?

Test/Inspection

1. Have the following tests/inspections been clearly detailed if required?
   a. bond (copper to laminate)
   b. cleavage (laminate to itself)
   c. dip solder
   d. hot oil
   e. dielectric K and disposition factor

   f. dielectric strength
   g. insulation resistance
   h. capacitance
   i. impedance
   j. vibration

   k. temperature cycling
   l. altitude
   m. humidity cycling
   n. flammability
   o. X-rays

2. Has temperature cycling between hot and cold extremes been used to expose marginal plated-through-hole and printed-circuit runs?
3. Have sampled plated-through-holes been microsectioned and examined? Have definitions been established for allowable holes, voids, or cavities? Under magnification, are coupons to show no holes, voids, or cavities? Can holes, voids, or cavities be irregularly shaped?
4. Are visual aids for accept/reject criteria and specified magnification utilized in inspection? Can undercut at through holes be contoured?
5. Have the flexure strength, insulation resistance, and high-potential (dielectric) tests been accomplished per applicable specs?

Additional Considerations for Multilayer P.C. Boards

1. Has a visual criteria been established for accept/reject of through holes? Both conducting and insulating holes?
2. What control is established over etch-back of insulation between conducting boards? Can fibre glass needles extend into solder?
3. Does it include examination of test coupons as described in paragraph 3.15 of item B-28-e?
4. How much "feathering" of electroless soldering is allowed in hole? How much buildup of "nodules" of electroless solder is acceptable? Under what magnification?
5. How is thickness and continuity of electroless nickel disposition checked?
6. What variation is allowed in contour of insulating material at holes after etching? Is paragraph 4.1.23 of item *B*-28-e used as criteria?
7. What control is established to prevent electroless nickel from bridging over air gaps in etched-backed areas?
8. What precautions are taken to maintain a cleanly cut hole? Are high speed carbide drills specified? When are drills replaced?
9. Have the applicable portions of specifications in paragraph *B*-28 been applied to drawing and specs?
10. Has criteria been established for accept/reject criteria by X-rays?

37. Radar Design Checklist

Eugene Nordell, Manager
Technical Evaluations and Management Systems
General Electric

This checklist on design and test requirements provides a generic method of aiding the development team in quickly attaining an overall perspective and definition of their problems and design objectives. Test functions and installation check-out requirements and procedures are readily derived from a similar checklist approach.

This checklist is intended as a reminder or guide to use in the preparation of a specific checklist for a radar system or related components. It provides the first step for iteration towards a specific checklist—tailored to the project at hand. By so doing, the necessary design perspective or vision will Simply, Effectively, and Efficiently (SEE) emanate. It must be kept in mind that these lists are only a point of departure which illustrate a technique, and require refinement and expansion for any application. This is best done by the most knowledgeable people—those to whom the design assignment is given.

Organization of the basic checklist is along the usual engineering family tree wherein trunk limbs and branches are replaced by system, subsystem, and component. Each subdivision of the family tree is categorized and includes some examples of environments, physical characteristics, and interfaces, as well as the functional requirements.

System

1. Function Description

   Has a functional description been made of the radar which describes the required function(s) per examples like those listed below?

   a. surveillance detection

b. acquisition designation
c. height finding
d. tracking discrimination
e. passive mode

f. frequency analysis, filtering, etc.
g. IFF (identification - friend, or foe)
h. chaff mapping
i. data recording
j. terrain avoidance

k. terrain mapping
l. storm mapping
m. instrument landing system (ILS)

2. Performance requirements
Note: *search, **track, √either
**a.* pd (probability of detection)
**b.* fa (false alarm rate)
√*c.* coverage envelope
range
AZ
EL
√*d.* resolution (range, angle, velocity)
Range
Min. - Meters (KM)
Max. - *h*-miles
Angle
AZ
EL-I degrees
Velocity
Min M/S
Max Km/s (knots)
**e.* dwell time
√*f.* data rate
√*g.* accuracy
h. clutter performance
i. sensitivity (min. dis. signal)
j. frequency
k. bandwidth

3. Target model(s)
a. effective echo area
b. glint characteristics
phase

amplitude
rate (time constant)

c. head-on echo area
d. tail-on echo area
e. distribution of target velocities

f. target scattering dist. (fixed, fluctuating)
g. target(s) spatial distribution
h. target echo areas frequency dependence
i. target aspect function
j. W/CDA, beta

4. *E*lectronic *C*ounter *M*easures (ECM)

What are expected levels in terms of:

a. frequency distributions
b. power levels
c. modulation types
  **c*ontinuous *w*ave (CW)
  **m*odulated *c*ontinuous *w*ave (MCW)
  *pulsed
  *noise
d. type and number of jammers
  *repeaters
  *sweep
  *CW, MCW
  *noise
    broad band
    narrow band
  *spot
  *barrage
e. spatial distribution?

5. *E*lectronic *C*ounter *C*ounter *M*easures (ECCM)

What is specified signal to jamming enhancement and allowable levels of degradation for: [specify jammer mix (types) and quantity]

a. data rate
b. maximum range
c. sensitivity
d. accuracy
e. automatic or manual ECCM mode selection?

6. Output requirements

   What is the format, accuracy, amplitude, and quantity of output data required, such as:

   a. visual displays, including both type and quantity, i.e. *A* - scope (2) *B* - scope (5), PP1 (10, etc.)
   b. telemetry
   c. automatic processing
   d. analog, plot board, synchro
   e. digital, encode
   f. data rate, levels, type, number of channels
   g. aural?

7. Special considerations

   Have the following (and others) been clearly and adequately defined in the specifications?

   a. special site requirements
   b. head end burn out protection
   c. cold start warm-up time
   d. blast hardness
   e. EMP (electro magnetic pulse) resistance

8. Environments

   Have all requirements been clearly defined for storage, transportation, and operation? Have the following been considered:

   a. radiation (level and type)
   b. shock
   c. vibration
   d. rain
   e. wind
   f. dust
   g. salt air?

   (See several environmental·checklists in Vol. VI and VII.)

9. Subsystem

   Have subsystems been adequately defined as to:

   a. functions
   b. interface definitions
   c. packaging
   d. dimensions
   e. weight, etc.

   Have the necessary design reviews been accomplished to insure optimum choice of techniques to satisfy functional requirements?

10. Interfaces

Have all functional, electrical, mechanical, and physical interfaces been defined and adequately characterized?

a. power input—defined in terms of demand profile, voltage fluctuation of rate, allowable harmonic content, power factor, etc.?

b. cooling/heating—including water and/or air flow rate, temperature gradient via connecting devices to external units, conduction, convention, radiation, etc.

c. housing/space—are allocations adequate by defined floor space, high/low bay, lighting, temperature control, floor loading constraints, etc.

d. functional—are all inputs and outputs established defined?

11. Environment

Considerations including, but not limited to, temperature, humidity, shock, vibration, radiation, humidity, etc., for various categories as use, transport, storage, internal, external, etc.

12. Physical characteristics

Defined to include number and type of enclosures, weight, layout, shelf life, appearance, etc.

13. Reliability: MTBF________hours at________per cent confidence.

14. Maintainability: MTTR____________ hours.

Subsystems

A great many subsystems are possible in today's increasingly complex radars. Only a few typical examples are included here. Included for each subsystem listed are a number of specifics, but implied is adequate consideration and definition of interfaces, physical characteristics (size, weight, etc.), and environment.

1. RF transmission

Subsystem includes all equipment necessary from the transmitter output to the radiating element (antenna). Antenna-drive command and control of rotating reflector systems, and the beam forming and steering function of a phased array would be exempted from this subsystem.

a. Voltage breakdown—considerations within the transmission line may require circulation of dry air, pressurized wave guide, etc.

b. Corona—prevention adequate, use choke/flange joints, specify finishes, plating, etc., on component parts. Wave guide switches and rotary joints are prime problem areas.

c. VSWR—(voltage standing wave ratio) adequately specified? Variations as function of frequency and antenna rotation specified? Is the rate of change vs. frequency compatible with the transmitter? Rotating joints can be trouble spots. Are isolators required?

d. TR Switch—Is recovery time specified in terms of range vs. attentuation?

2. Beam Forming

Subsystem considerations will vary, dependent upon use of a phased array, reflector, etc.

a. Are beam requirements adequately defined in terms of quantity, shape, (vertical and horizontal); fixed versus variable; special characteristics such as MTI "bump" in surveillance applications to compensate for range reduction at upper angles of coverage caused by clutter return. Is orientation of beam specified for a given coverage (i.e. lower 30 db. point on horizon). Are allowable limits specified over the entire required frequency band?

b. Error tolerance budget—applied to all component parts of the beam forming system if phased array is used? Separate phase and amplitude characteristics defined? Presence of ice, snow, etc., on external elements taken into account?

c. Scanning or beam positioning—requirements include possible wind effects on antenna drive, structure deformation, thermally induced changes, gear train and load stiffness and inertia, back lash, load resonance, etc.

3. Transmitter

Subsystem may include all reference frequencies generation up through power amplification to level necessary for driving the antenna, mixers, etc.

a. Frequency generator requirements complete in terms of quantity of simultaneous signals, range of tuning or discrete steps required, Intrapulse, Interpulse, and longer term stabilities defined in amplitude, phase, and rate terms? Frequency shift requirements documented and effect on other subsystems analyzed? Tradeoff of multiplier chain vs. high-level oscillator complete? Are stabilities and references adequately defined for signal processing subsystem (establish in FM deviation terms)?

b. Power level requirements defined in peak, average, duty cycle, code rate, and other data pertinent to a power *pulse*? Are there dual operating modes, and is the power measurement technique established and agreed upon?

c. Modulation of the base RF frequency specified in terms of pulse repetition rate, allowable envelope jitter, time delay from master trigger source to RF Pulse out? Are there dual modes required, i.e. different P.R.R. and pulse width for "burn-through", etc.?

d. Wave form (i.e. chirp, linear format of pulse train), defined: Pulse code, phase code, type mod of HPA —grid, mod anode, cathode.

e. Cooling requirements specified and heat exchange medium tradeoffs made of air vs. liquid, types of liquid, etc.?

f. Maintenance and replacement provisions for heavy components such as pulse transformers as magnetron or klystron final. Inattention to this one can quickly gain notoriety for your design!

g. Shielding for radiation including X-ray as well as RF hash and interference. An early test program may be desirable to substantiate analysis in this area. Measurement techniques for use in a validating test program warrant thorough study!

h. Safety provisions can be rendered temporarily inoperable during maintenance procedures, etc. Are safety requirements specified to include effectiveness during other than normal operating modes?

4. Receiver

Subsystem generally includes signal processing and ECCM processing as well. A major concern always is the extreme variations in signal level present, ranging from the nearby transmitter to the relatively miniscule target echo. Meticulous attention to power supply decoupling (conducted interference), shielding, and grounding networks is a must!

a. Noise figure able to survive front end overloading and "blast" from adjacent radars. Has overload limiter been considered?

b. Dynamic range adequately specified to include required responses such as linear and clip, linear plus logarithmic hard limit, etc.

c. Band pass characteristic stability vs. gain setting, signal level, temperature drift, etc.

d. Oscillations or ringing at low levels are quite often nondetectable except by their influence on precise signal processing equipment. Test programs, after exercising normal design precautions, are advantageous. Signal induced video current pulses (nonlinear operation), can induce "noncoherent" IF "ringing" or transients which may severely limit MTI cancellation ratios or coherent integration capability.

e. STC

f. AFC

g. Beacon/Sicin Option

h. Image rejection

i. AGC

5. Additional Subsystems could include:

a. display

b. beam steering

c. power distribution

d. trouble shooting

e. etc.

The same rationale of driving out all requirements and problem areas would of course apply.

Components

A component is simply a functional subdivision of a larger entity such as a subsystem. Included items would be, for example: IF amplifier, band pass filter, keep alive power supply, transmission line filter, etc.

1. Transmission line filter, special problem areas:

a. Material temperature coefficients compatible with operating temperature range and allowable variation in operating characteristics. Is temperature transient response acceptable?

b. Operating life affected by surface corrosion, etc., and attendant change in operating characteristics.

c. Is conductivity requirement sufficiently high to warrant drawn (polished) silver/gold finishes vs. deposited layer?

d. Physical machining tolerances and *progression of tool bit wear* are often vital success factors.

e. Band pass characteristics specified in terms of phase and amplitude response across sufficient frequency range?

f. Are cost targets established? In phased arrays, any precision multiple use component is a cost problem.

g. Is performance susceptible to installation procedures? Are guide pins, torque wrenches, etc. required?

h. What storage and handling safeguards are required?

38. Solid State Digital Component Checklist

R. DiNolti
Quality Control and Test Engineer
General Electric
Re-entry Systems Department

The problem of transition of many older electrical engineers from vacuum tube technology to solid state technology is a very real concern to many managers, particularly on digital devices.

Despite literally hundreds of documents to explain and train the application and functions of solid state devices, there has been a real need for a short, concise document which provides the technical review of specific items in semiconductor, electronic circuit design and analysis.

One of the documents which has reasonable success in this area is this "Solid State Digital Component Checklist." It has been of particular assistance to the ancillary groups in manufacturing, quality, and process control which do not have direct design experience responsibility.

This checklist was prepared for a specific digital component, such as clock circuit, or PCM multicoder. It is equally applicable to flip-flop, logic, gating, and other low-signal repetitive pulse network devices.

General

1. Does design satisfy all specification requirements?
2. Does design utilize standard, qualified, logic elements where possible? Has selection of special parts been checked?
3. Have component parts been sufficiently derated? By what rules?
4. Has a worst-case circuit analysis been performed on component? Are results available for review?
5. Have the EMI requirements been taken into consideration? (See Checklist 32.)

6. Has power line filtering been provided? Is noise level determined by bit resolution of system?
7. Will component satisfy specification requirements over entire input power frequency and voltage variation range?
8. Does component utilize a self-contained power source?
9. If so, has sufficient short circuit protection been provided? Will failure of one logic element destroy the power source?
10. If a self-contained supply, d.c.-to d.c. converter, or a.c.-to-d.c. converter, is used, is power for logic elements used for any other purpose (i.e. relay coil power)? If it is, has sufficient filtering been provided to maintain logic noise immunity?

11. Is the self-contained power source input sufficiently protected against power line transients? Spikes?
12. Has sufficient logic protection been provided so that a failed part doesn't completely disable component function (i.e. redundancy)?
13. Have "select by test" parts been held to a minimum and utilized only as an exception rather than a general rule?
14. Have environmental requirements been taken into consideration in the component design?
    a. Will the temperature change characteristics of the logic elements cause logic hang-ups or races?
    b. Will temperature change cause frequency drifts of internal clocks?
15. Have sufficient test points been brought out so that a comprehensive operational check may be given the component? (See Checklist 31 and checklists in Vol. V and VI.)

16. Has logic initialization been considered, i.e. is it possible for key logic elements to assume an initial state when power is applied that could result in system lockup or improper functioning?
17. Has a "system reset" function been provided in the event item 16, above, cannot be avoided?
18. Has consideration been given to the characteristic impedance of high-speed clock lines? Can mismatch cause standing waves along conductor? If so, can negative overshoot be tolerated on logic element inputs?
19. Have problems associated with interfacing synchronous signals been considered, i.e. the occurrence of an external event (random) in relation to system clock?

20. Is a power-up/power-down sequence required? How are logic circuits reinitialized in event of a power failure/recovery?

21. Does the design rely on device propagation delays for proper operation? If so, have minimum as well as maximum delays been considered? How do delays vary over temperature range?
22. Has the design considered ways to test the device and/or isolate troubles? If operation of the device depends on the response of an external device-to-device request, is there a method of simulating this interface easily in order to isolate faults? This requirement implies more than just adequate test points; a means must be devised to enter a nonoperating "closed-loop system," and to simulate subsystem responses.

Example (PCM Multicoder)

1. Does multicoder satisfy system, subsystem, and component requirements with regards to accuracy, bit resolution, etc.?
2. Has the choice of logic been determined by subsystem design, or is it due to available elements?
3. Is input and analog data high level or low level? Is it mixed?
4. Is input and signal conditioning required? Are inputs direct or time division multiplexed?
5. Is input data analog, digital, or both?

6. Is data output serial or parallel?
7. Is the component resolution and accuracy flexible enough so that a requirement change necessitates minimum redesign? Are spare inputs available?
8. What is output format? NRZ, RZ, Manchester code? Is it compatible with ground station?
9. Have positive sync, frame, and subframe words been incorporated? Is sync word one that could not be reproduced by a data input?
10. Is output data signal conditioning required? Is pulse shaping required to deviate a transmitter or drive a recorder, etc.?

11. Is an internal power source used, or is line voltage conditioned and used?
12. Has the internal source been designed for peak efficiency? Is it a requirement?

13. Is the power source provided with short circuit projection? Will a shorted logic element destroy the power source?
14. Is power source noise-free? Has sufficient filtering been provided to keep noise down? What is allowable noise level?
15. Has internal noise been accounted for?

16. Have logic circuits been designed or selected with high noise immunity? Has general practice of keeping the noise immunity between 10 and 20 per cent of the logic signal swing been used? What is allowable "jitter"?
17. Has temperature vs. noise immunity been taken into account?
18. Has the mutual capacitance of interconnecting conductors been taken into account? Are interconnecting wires carefully routed and laced or is a "rat's nest" point-to-point wiring used?
19. Are the signal lines most susceptible to mutual or stray capacity kept as short as possible?
20. Have twisted shielded pairs been used where needed? If so, has circuit drive been considered? Is buffering needed?

21. Have decoupling capacitors been provided at regular intervals along the power lines?
22. Have ground loops been eliminated by design? Have grounds been kept short where possible to keep ground noise down?
23. What type internal clock is used? Free-running multivibrator or crystal controlled oscillator?
24. If free-running multi, has a sure start been provided to eliminate hang-up? Has buffering been provided?
25. If crystal oscillator is used, has crystal been sufficiently protected against handling, shocks, hard environments, etc.?

26. How is crystal shocked into oscillation? Is it fail-safe?
27. Is crystal stable enough to provide accuracy? Has oscillator been designed to protect against moding?
28. Is the clock stable over required temperature ranges? Over load variations? Over combined temperature, load and input variations?
29. Is clock resync required, or is it free-running? If external sync is used, has sufficient filtering and buffering been provided?

39. Systems Design Checklist

N. Haley
Grumman Aircraft Engineering
Company

When all is said and done about the factors affecting component and part selection, the real point of maximum concern is that the systems design be adequate.

By systems design is implied all the customer requirements as well as the proper interfacing with other systems, with the subsystem within the system, and the multitude of test situations in-process, final, and in use by the customer.

Far too often can be seen the results of the complacency, näiveté or carelessness of top-level systems design personnel. These circumstances arise for many reasons and from many pressures, but the specific use of a good systems design checklist is at the systems interface meetings, the system integration meetings, and the design reviews, which can greatly minimize the inevitable failures and excessive costs that can result from shortcutting these steps.

1. Have the input-output impedances been properly expressed and analyzed with interfacing impedances?
2. Will the design operate correctly with adequate margin for uncertainties in the specified supply voltages?
3. Are adequate features provided to protect circuits against excessive variations in line voltage?
4. Are adequate features provided to protect circuits against inadvertent loss of line voltage and/or $B+$ and/or bias voltage(s)?
5. Is circuit grounding correct and in accordance with overall grounding mandates?
6. Is a.c. grounding adequate to stabilize circuit ground relative to chassis ground?

7. Are the interference limits known? Injected? Generated? (See also Checklist 32.)
8. If so, do the circuits have adequate design features to provide an adequate operating margin?
9. If interference limits are unknown, has an analysis been made to ascertain the worst-case, and does circuit provide margin for it?
10. Is overall protection provided against overload, short circuit, excessive heating, pressure changes, etc.?

11. Do self-test features meet the applicable requirements?
12. Have "preferred circuits" been used whenever possible?
13. What alternate circuits or systems were considered?
14. What factors influenced the choice of this particular circuit or system?
15. Are the specifications for this circuit firmed up, including test specifications?

16. Has every effort been made to relax any unreasonable or unusually difficult performance or environmental requirements?
17. Have phase-margin checks been performed on all feedback loops?
18. What decoupling or neutralization schemes have been implemented to avoid regenerative feedback loops?
19. What analyses were conducted to determine the existence of feedback loops and their effects on other circuits?
20. What effect does length of input/output cables have on performance?

21. Can any circuits be simplified or eliminated and still allow the system to operate within the requirements? (See "Human Factors Checklist" in Vol. V.)
22. Can the system or circuit operate satisfactorily after the minimum required warm-up time?
23. What system or component adjustments are required when a module or unit is replaced?
24. What design features were incorporated to suppress parasitic oscillations?
25. Have appropriate test points been established for troubleshooting?

26. Do tests show functional compatibilities of the subsystem and system when operated in conjunction with interfacing operational and/or test equipment?
27. Are the specifications for the circuit, including the test specifications, written in approved format?
28. Does the test specification clearly and concisely note the procedure to be followed in operation of the test equipment ?
29. Can the test specification be adequately understood by the lowest experience level of test operators?
30. Has circuit redundancy been used where required? Have adequate tests been made to verify redundancy feature?

31. Will undesirable coupling occur through the power supply?
32. Are the load factors known? If the system flies in an aircraft, is it compatible with MIL-STD-704?
33. Does the output stay within tolerance under conditions, singly or in combination, of:
    a. minimum to maximum variation of input signal parameter(s)
    b. permissible extremes of input power regulation, voltage
    c. extremes of operational environment (temperature, altitude, humidity, vibration, etc.)
    d. end life characteristics of parts (degraded values due to age and drift)?
34. Can system "start" under full load? Is this a requirement?
35. Have incompatible tolerance requirements in various levels of tests been prevented? Are component test tolerances tighter, looser, or equal to system test requirements?

36. Is "select by test" of components or parts necessary?
37. Has consideration been given to a commercially available test unit in lieu of an internally designed and developed unit?
38. Can the specified test equipment adequately measure the extreme low and high parametric readouts required? Are nanosecond time measurements required?
39. Are the test requirements based on real parameters?
40. Will the test equipment actually indicate a "go" or "no-go" condition which can be used as a basis for acceptance or rejection of parts?

41. Do specifications include figures in polar or rectangular form of input and output impedance, where applicable?

42. Do specifications include figures for maximum allowable deviation from the ideal wave form (for sine waves distortion in percent of harmonic content, for pulses, rise and fall time)?
43. Do specifications include figures of maximum allowable noise levels? On a.c. and d.c. inputs? On signal input? On output?
44. Has a fail-safe design philosophy been used?
45. Have the applicable reliability specifications been reviewed to determine the scope and type of data that will be required for reliability reports? (See checklists in Vol. V.)
46. Has "Maintainability Checklist" in Vol. V been reviewed?
47. Do test engineers know the difference between precision, accuracy, repeatability?
48. Is interaction between adjustments and other circuit parameters minimized?
49. Does the design preclude damage to the circuit by the careless use of an adjustment or combination of adjustments? If not, have adequate precautions been provided for same in specification, manuals, etc.?
50. Is the necessity for, and frequency of, periodic testing specified?

40. Transformer and Inductor Checklist

This checklist was prepared by a company which pioneered the use of a computer to solve the design problems of size, weight, power, impedance, wire size, efficiency, IR loses, and other factors.

As part of the program to review an engineer's inputs into both the computer and to drafting, and for later review by manufacturing and quality, this checklist was prepared. It is equally useful to any company that winds coils or makes inductive devices even if they have not reached the level of sophistication where a computer is utilized to solve design equations.

This checklist is an excellent example of the type of detailed checklist which can and should be a part of the everyday procedure in preparing a design, drawing, specification, planning procedure, or new product.

1. Inductance
   - *a.* What is standard tolerance?
   - *b.* What is standard test voltage? a.c. or d.c. or both?
   - *c.* What is test frequency?
   - *d.* What are details of taps and tolerance?
2.. Is maximum d.c. resistance specified?
   - *a.* Since spools of wire vary and the permeability of cores varies requiring turns adjustment, this is better than a ± tolerance on DCR.
   - *b.* Is test method and equipment specified? Measuring DCR on some cores applies a polarizing voltage detrimental to the future use of the component.
3. What is minimum self-resonant frequency? What test method and equipment?
4. What is maximum distributed capacity, coil-to-coil and coils-to-case? What test method and equipment?
5. What is maximum working voltage level a d.c. component will be required to handle?

6. What is allowable decrease in inductance with applied voltage level or direct current?
7. Is Q not less than specified value(s) at discrete frequency(ies)?
8. What is form factor?
    a. Envelope dimensions, tolerances
    b. lead location, tolerances
9. Are lead details provided?
    a. gage, tolerances
    b. length, tolerances, or a minimum material
    c. stripping, tinning
10. Are marking details provided?
    a. maintenance name or identification number
    b. manufacturers' part number and revision
    c. customer part number and revision
    d. value of inductance, tolerance
    e. date code
11. Are applicable portions of the following MIL specs applied?
12. Are extreme environmental requirements detailed? (See environmental checklist in Vols. VI and VII.)
    a. shock
    b. humidity
    c. vibration
    d. temperature range
13. Are insulation and wrapping details provided? Winding-to-case?

41. Transmitter Design Checklist

Courtesy of Microcom

This checklist was prepared for frequency and phase modulated aircraft, missile, and space type transmitters operating in the *S* and *X* bands. It represents the kind of detailed practical design checklists which a design engineer would not mind reviewing with his peers, be they in reliability, quality control, or systems test. All the questions are relevant and vital to assuring full consideration has been given to all specification requirements.

Input Power

1. Are the input power requirements of the impedance, transmitter fully defined in terms of:
   - *a.* voltage
   - *b.* current
   - *c.* noise
   - *d.* transients
   - *e.* spikes
   - *f.* regulation
   - *g.* interrupted power?
2. Has a tradeoff been made between the d.c. isolation gained by floating the *B*—return vs. the large capacity resulting between true ground and the *B*—return?

Oscillator

1. Will oscillator(s) remain within spec limits over temperature range? During vibration? After mechanical shock?
2. Is the case sufficiently rigid to prevent oscillator shift during normal screw-down procedure to proper mounting surface?

3. Are critical oscillator component parts aged or burned in where applicable?
4. Will curing or aging of foam, potting, or encapsulants cause mechanical pressures affecting oscillator outputs? This item is a question for the system designer and the transmitter designer.

Modulation

1. Can any parameter of input modulation cause:
   a. distortion
   b. over modulation
   c. clipping
   d. unwanted intermodulation products?

RFI (See Checklist 32.)

1. Is it *necessary* to design to all the requirements of MIL-STD-461A, MIL-STD-462, or MIL-STD-463, or other specs? Large costs can be eliminated by applying only absolutely necessary requirements. For example: launching operations usually only require frequency stability and antenna conducted spurious signals to meet EMI/RFI specs.
2. Is an overall RFI check made to determine if the transmitter generates and injects conducted or radiated interference into the input lines?
3. Is an overall audio susceptibility test made? Are susceptibility voltages of specified input runs compatible with impedance, power, and application?
4. Can electrical noise cause objectionable carrier modulation?

Output Power

1. Are the output power requirements of the transmitter fully defined in terms of:
   a. impedance
   b. voltage
   c. current
   d. distortion
   e. noise
   f. modulation?
2. Does transmitter operate at maximum power at maximum temperature? If not, at what environmental temperature?

3. Does the mounting configuration provide a satisfactory heat sink to prevent electrical degradation?
4. What test load is to be used when antenna and/or coaxial lead is unavailable?
5. Is power measured with or without modulation? By what method and why?

6. Will modulation and distortion be measured with an approved receiver and distortion analyzer?
7. What symmetry will be required in modulation of side bands?
8. Again, the above is a question for the system designer. The transmitter designer is only responsible for meeting the specified power output at the output connector of the transmitter. What happens after that is the responsibility of the system designer.

VSWR (Voltage Standing Wave Ratio)

1. Will the transmitter operate satisfactorily after:
    a. open circuit antenna (specify cable length)
    b. short circuit antenna (specify cable length)
    c. no breakup with 2/1, 5/1 mismatch at all phase angles
    d. minimum and maximum radiated power with 2/1 and 5/1 mismatch
    e. spurious signal outputs with 2/1 and 5/1 mismatch at all phase angles?
2. What variation in antenna impedance can occur before the transmitter performance falls out of spec?
3. Has the performance degradation with antenna mismatch been calculated assuming a fixed transmitter output Z?
4. Is an actual test made with through-line watt meters and the Smith charts?
5. Is an isolator incorporated in design? In test?

6. Will "breakup" (undesirable oscillations in output circuit) occur from reflected Z back into output stage?
7. Will excessive or damaging B+ currents result from detunning caused by reflected Z?

Environmental (See also checklists in Vols. VI and VII.)

1. Are the transmitter's performance characteristics and limitations fully defined under the extremes of electrical and en-

vironmental conditions in its spec? In its end use application?

2. Can vibration cause objectionable carrier indulation?
3. In most cases of *S*-band design, a cavity filter is required where large RF voltages are developed. Most *S*-band transmitters generating over 1 watt of power and using a 2 or 3 section cavity filter will break down under vacuum unless special precautions are taken. Many designs require pressurization. This item should be included under an environmental section to ensure the designer considered the problems of transmitting through the critical altitude.
4. Will cavity filters require special precautions to pass thermal/vacuum tests?
5. Will part of transmitter be pressurized to assure proper transmission at critical altitudes?

6. The above should be part of a standard specification on frequency stability under all environments. This should be questioned under the electrical specification section of the checklist, and the unit either meets the frequency stability requirements of the specification or it does not.

Mechanical (See Mechanical Checklists in Vol. III.)

42. Transponder/Beacon Design Checklist

Joe Hancox
Quality Control and Test Engineer
Re-Entry Systems Dept.
General Electric

One of the most critical components manufactured in today's aircraft and missile industry is the transponder or beacon. It receives radar signals from primarily ground or ship based radar, interrogates the signal and, thus keyed, transmits an identifying pulse back to a radar station. With increased concern over aircraft collisions, and with the newest jumbo aircraft now on the scene, this component becomes more and more common.

This checklist is indicative of the type which can be utilized in product planning, specification writing, design study, design reviews, and in customer discussions, including test, quality, and reliability factors.

Electrical

1. Will this design meet its electromagnetic interference (EMI) requirements?
2. Does the conducted interference test data exhibit sufficient, or only marginal, limits?
3. Is RF radiated interference well within its db limits over the required spectrum?
4. Are the RF antenna-conducted interferences within acceptable limits?
5. Does the data indicate an acceptable RF conducted and AF conducted susceptibility margin?
6. Is the transient conducted within the particular design limits?
7. Which EMI/RFI spec is applicable? (See Checklist 32.)
8. Are all performance parameters within the desired design performance range?

9. Will all parameters meet the minimum requirements throughout upper- and lower-line voltage variation limits? And environment combinations?
10. Are minimum and maximum line current requirements maintained for environmental and performance extremes?

11. Is the required "interrogation code" capable of the variety of interrogation codes specified?
12. What is stability (tolerance) of accept and reject zones? Is the margin adequate?
    a. Has there been a sufficient interrogation window slot analysis?
    b. What happens if the slot shifts from the set limits?
13. Is the system sensitivity met? What is the minimum acceptable (dbm) signal?
14. Was the average sensitivity obtained from a representative sample?
15. Is the receiver frequency stability adequate? Was the original design tolerance met?

16. What frequency drift compromises exist?
17. Does the receiver bandwidth meet the minimum acceptable limit? Is it typical of a representative sample?
18. Is the receiver selectivity marginal? Is a conventional type frequency conversion utilized?
19. Does image rejection indicate a sufficient margin?
20. What is the typical power output? Does it approach the maximum or minimum limit?

21. How stable is the transmitter frequency? What are the drift limits?
22. Is the desired design transmitter response pulse shape maintained?
23. What is the rise time, pulse width, fall time?
24. How good is the transmitter bandwidth? Does it approach the upper or lower limit?
25. What spurious signal problems have appeared in testing to date? Is there an RF filter employed? Should there be?

26. Does the fixed delay fall within the limits? Should it be variable?
27. Is the delay variation adequate? What is the margin?
28. Is the delay jitter within the required limits or are best achieved results acceptable?

29. Does this design appropriate transponder/beacon protection/safety features?
30. Are circuits terminated into the specified impedances?

31. Is the over-interrogation (P.P.S.) provision adequate?
32. Does adjustment of the over-interrogation range limits affect performance?
33. Is the input voltage reversal protection infallible?
34. What is the effect of an external short across the telemetry circuit? Does the condition permanently affect circuit performance?
35. Does an antenna mismatch (or short circuit) permanently degrade antenna performance? Or is protection provided only for a degree of mismatch?

36. How much receiver input power can the mixer crystal withstand?
37. Does the crystal mixer breadboard data indicate an adequate margin for the design requirements? How much power (dbm) can the receiver withstand?
38. Does the modulator design preclude modulator "hang-up"? What are modulation limits?
39. If preventative modulator "hang-up" is incorporated, will subsequent operation of this circuitry affect performance?
40. Is modulator recovery time adequate?

41. Does this design employ cold firing protection? Is this elapsed time adequate?
42. Does this design incorporate sufficient telemetry monitoring circuits?
43. Have the telemetry monitor outputs been properly isolated and terminated?
44. Are the telemetry output signals compatible with the external associated circuits?
45. Do the telemetry circuits affect performance?

46. Are all electrical parameters met?
47. Has the basic design philosophy been adhered to?
48. Is there anticipated any significant performance degradation as a result of packaging? When lid is closed?
49. Is packaging density reasonable? Are modules easily manufactured, easily tested?
50. Are the TBD (to be determined) part call-outs kept to a minimum?

51. Are Hi-Rel and Mil-Spec parts called out as required?
52. Are Hi-Rel parts called for when Mil-Spec parts could be used consistent with component application and criticalness?
53. Have all parts been properly derated for their circuit application?
54. Have all critical subassemblies been properly preconditioned? Aged or burned in?
55. Have all factors of Product Assurance Checklists in Vol. V been considered for "ease of adjustment, testing and alignment"?
56. Have voltage breakdown, corona, and arcing in the RF power areas been considered?
57. Are there sufficient adjustment features available?
58. What special attention should be paid to the total input-output function, power, frequency, etc.?
59. Are there special parts or components in this design requiring special handling or testing?
60. Is documentation necessary to control special parts/components?
61. Has appropriate consideration been given to the electrical design with respect to the environmental requirements, i.e. temperature, shock, vibration, acceleration, etc.?
62. Was adequate breadboard testing performed?
63. Does test data reveal that more development work should be undertaken?

Mechanical

1. Have all requirements pertaining to dissimilar metals, finishes, nonfungus nutrient materials, etc., been considered?
2. Does the design meet the desired construction size, weight, outline, and configuration?
3. Is the design adequate with respect to the stresses encountered as a result of environmental conditions, i.e. temperature, shock, vibration, acceleration, pressure, etc.?
4. What is the allowable leakage rate? Is hermetic seal or a pressure seal required? Is the seal adequate or marginal?
5. Are all provisions for mounting and mechanical mating compatible?

Chapter 4

MICROELECTRONIC CHECKLISTS

43. Integrated Circuit Specification Checklist

Ed O'Connell
Rome Air Development Center
Griffiss Air Force Base
New York

When it comes to establishing the specific characteristics, test, quality/reliability requirement for integrated circuits (ICs), the drawing specifications may frequently contain as many as seventy-five pages.

To assure that all general and quality/reliability requirements would be considered (but not necessarily applied) for military or commercial microcircuits, and to present them in understandable terms and in an acceptable format, the government released a new specification, MIL-M-38510 (Military Specification Micro Circuits, General Specification for), early in 1970. Also scheduled for release then was the EIA (Electronic Industries Association) General Specification for Microelectronic Devices, known as the MED (Microelectronic Devices) Tentative Standard 3.

This thirty-page document covers the factors which need to be resolved as part of definitizing the procurement package. The ordering of ICs can be critical, and can be significantly helped by the use of a checklist such as this one. It was prepared by the reliability branch at the USAF Rome Air Development Center where the original MIL-STD-883 was developed, and where the final draft of MIL-M-38510 was released.

Outline Drawing Requirements

1. Are top, end, and side views (where required) of the device package drawn to scale?
2. Does it include all mounting lead and case dimensions (with tolerances)?
3. Does it indicate controlled dimensions of leads? Uncontrolled dimensions of leads approaching the case?
4. On the device leads:
   a. Is quantity of leads specified?
   b. Does it specify if leads are to be isolated from case?
   c. Does it specify what to do about missing leads?
   d. Does it specify what to do about unused leads?
5. Does it indicate with dots or special lead marking the sequence (index points) of numbering the leads?
6. If part is claimed to be a standard part, does it have an EIA (Electronics Industries Association) registration number?
7. Are all dimensions compatible with EIA requirements, including all applicable dimensions being referenced to the mounting plane?
8. Is a fully descriptive name of device included on drawing?
9. Does it detail all marking requirements, including P/N, serialization, location, and permanency of the marking? (See paragraph 3.6 of MIL-M-38510.)
10. Has Parts Engineering checked to assure the following are included in the detail drawing specification:
    a. part number (including date and revision letter)
    b. spec number (including date and revision letter)
    c. design, construction, and physical dimensions
    d. special marking provisions
    e. acceptance numbers for AQL, LTPD, or PDA if applicable
    f. identification of electrical parameters, limits, and conditions for acceptance test
    g. package thermal requirements?

11. Have Purchasing and QC checked the following ordering data:
    - *a.* All items in #10 above
    - *b.* If test data is to be furnished
    - *c.* Are product assurance test and inspection levels understood
    - *d.* What design documentation is to be furnished
    - *e.* Are welding or other special process data or test coupons required to be submitted
    - *f.* Who does what and when on failure analyses
    - *g.* Are requirements governing changes in design, test, or processes understood
    - *h.* What reliability demonstrations are required
    - *i.* Are lot controls practical
    - *j.* Can a standard part be used?
12. Are all special end item environments, which the device must withstand in use, included in an applicable portion of test program? For example, if high pressure or radiation hardened devices are required, are tests included in specification? See the section, "Environmental Test Requirements," in this checklist.

Configuration Requirements (See also "Configuration Management Checklist" in Vol. I.)

1. Do drawing blocks contain signatures for draftsman, checker, design engineer, quality/reliability engineer?
2. Do drawings meet the requirements of MIL-D-1000 (or MIL-D-70327) and MIL-STD-100?
3. Are changes to each page of drawing and specification controlled per ANA Bulletin 445 (or the newer MIL-STD-481 and 482)?
4. Are adequate details provided which will control functional interchangeability and pin-to-pin replaceability? For any device supplied in conformance with detail specification and part number?
5. See checklist for MIL-D-1000 and MIL-D-70327 in Vol. VI.

Mechanical and Thermal Requirements

1. Are requirements of case, glass insulator beads, and leads clearly established in spec for:
    - *a.* fungus resistance

b. corrosion resistance
c. lead material/heat treat condition/surface condition
d. lead forming (if required)?

2. Are the method of sealing and requirements for inspection/test of same clearly specified? See "Hermetic Sealing and Leakage Checklist" in Vol. IV.
3. Is the extent, type, method, and *permanency* of marking of device clearly defined?
4. Is operating, storage, and maximum/minimum temperature range specified?
5. Are detailed workmanship requirements included? Do these make ample use of visual aids and contain accept/reject criteria? *Is a precap visual inspection required per Method 2010 of MIL-STD-883?*
6. Is a note included which contractually restricts supplier from making *major* process changes without prior discussion with buyer?
7. Is delivery packaging in accordance with MIL-STD-1276 and/or MIL-M-55565? Is the specific packing container indicated on the part drawing? It should be!
8. Are preservation requirements specified to level *A, B,* or *C* of MIL-M-55565?
9. Are requirements for gross and fine leak test detailed in specification? Or is Method 1014 of MIL-STD-883 referenced?
10. Have all the requirements of the "Environmental Test Requirements" section of this checklist been considered?
11. Since solderability can be a major problem, what specific requirements have been included in the specification for the particular type of soldering (dip, hand, wave) which will be used?
    a. MIL-STD-750
    b. MIL-STD-883
    c. Other (See also "Soldering Checklist" in Vol. VIII.)
12. Is tin plating of leads acceptable? Or is gold plating required? If gold plating, is tin to be applied over the gold?
13. If gold plating is required, is it specified per MIL-G-45204 or equivalent specification? Is type I or II specified, since that can affect solder wetting temperatures?
14. Are any special assembly processes detailed in specification? Items such as:
    a. all aluminum construction

b. special die attachments
c. bond integrity.

15. Are case material and plating thickness defined?

16. Are the following defined:
    a. wire bonding method
    b. bonding wire material
    c. wire bond pull test (If so, at what frequency and sample size?)
    d. die bond test
    e. die mount technique, cold or hot?

Electrical Requirements

1 Are all the electrical parameters specified at the appropriate temperature ratings (low, ambient, and high)?
2. Have separate temperature limits been established for storage/transportation?
3. Do the electrical parameters include as a minimum the following (as applicable for each section or gate):
    a. input voltage(s) (continuous or pulsed)
    b. input frequency, pulse, transient, ripple, spike, or other limitations
    c. input forward current
    d. input reverse current
    e. output voltage(s)
    f. short circuit current
    g. output leakage current
    h. switching times, rise and fall times
    i. power drain?
4. Or are the parameters selected from tables in MIL-STD-1331?
    a. Table I — Dynamic
    b. Table II — Static
    c. Table III — Amplifier
5. Is a logic and/or schematic diagram included in which all inputs and outputs are clearly and completely labeled? Is each function clearly identified with its own inputs and outputs?
6. Is test schematic and output load circuit shown for all lot acceptance and burn-in testing?
7. Is a schematic and test circuit and instrumentation shown for any switching time tests? Are wave shapes shown when ne-

cessary? Are test lead lengths and types indicated where necessary? See Method 1015 for five different test conditions.

8. Has a matrix been made to indicate every applicable test current/voltage and characteristic for each pin under test? Is ground connection(s) clearly indicated? Are symbols clearly understood?
9. Is testing accomplished using MIL-STD-883 procedures?
10. Are maximum electrical ratings clearly defined? At what temperature?
11. Is a reverse bias burn-in test, per Method 1015 of MIL-STD-883, to be run on all units or only Class *A* devices? Will devices be returned to room temperature with bias voltage(s) maintained?
12. Have all the environments of Section *F*, paragraph 1, been specifically reviewed to assure selection of the proper method number?
13. What values have been allowed for burn-in testing?
14. Is stability (temperature coefficient) required on a lot basis?

Quality/Reliability/Requirements

1. Are all quality/test/reliability requirements included in the purchase order package? Is device class specifically identified?
2. What traceability requirements have been applied on supplier?
3. Is GSI (Government Source Inspection) required? Is customer final and/or in-process inspection required?
4. Is customer surveillance inspection necessary? Desirable? Not necessary?
5. If necessary, does it include any phase of manufacturing flow chart not designated as proprietory?
6. Will all test equipment be calibrated per MIL-C-45662A?
7. Will data sheets contain as a minimum (or equivalent automatic print outs):
   a. drawing number and suffix (rev.) number
   b. device serial number
   c. lot code number
   d. tests and environments
   e. decimal locator key and unit?

   Will data sheets be approved and signed by suppliers QC?

8. Will supplier provide a complete process flow chart referencing the major production processes, controls, fixtures, test procedures, etc. (See Flow Chart Checklist, Vol. VI.)
9. Will any test procedures, test data, pertinent documents, etc. be furnished upon request? If an exception is made to proprietary operations, have these areas been defined in the purchase order package? Why not?
10. Will country of origin be indicated on paperwork accompanying the devices?

11. What lot identification is required? How is lot defined? Does this definition cover photo master, wafer diffusion, assembly lot, test lot? How is identity marked on I.C.s and on software?
12. Which provisions of MIL-M-38510 (Method 5005 of MIL-STD-883) or EIA MED Tentative Standard 3 applied to qualification by similarity will be accepted?
13. What acceptance tests/inspection are to be 100 per cent? Which are to be according to MIL-STD-105D (specify AQL, normal inspection, single sampling) or LTPD (lot tolerance percent defective) *with maximum accept number specified*? Or will AOQL be utilized?
14. Will a 100 per cent precap inspection (per MIL-STD-883, Method 2010, Test Condition *a* or *b*) be required to be witnessed by customer representative?
15. What specific 100 per cent screening/inspections/tests are required? Preconditioning and screening? *Note*: If order is important, is it specified? Remember, the last test is to be the Final Electrical Test.
    - *a.* temperature stabilization
    - *b.* temperature cycling (nonoperating)
    - *c.* constant acceleration
    - *d.* hermetic seal (gross leak)
    - *e.* hermetic seal (fine leak)

    - *f.* burn-in test [Does PDA (percent defective allowable) need to be applied?]
    - *g.* radiographic (X-ray)
    - *h.* temperature/shock/vibration

16. Is supplier required to submit a reliability, inspection and/or test plan? If so, is it per MIL-STD-790? Are sample data sheets included in it?

17. When will plans in item 16 be approved or rejected by purchaser?
18. What procedures and controls are required on rework of units at supplier? For example, is rebonding permitted? Are these items clarified in the spec?
19. What controls are established on test method deviations? See paragraph 4.3.4 of MIL-M-38510.
20. Are supplier's inspection and detail test plan and procedures subject to customer approval?

21. Are test operator errors or equipment malfunctions controlled per paragraph 4.3.5 of MIL-M-38510?
22. Is test data indicated as to which parameters are recorded by attribute and which by variables data techniques? Which are to be permanently recorded?
23. Are procedure and data to demonstrate initial qualification per paragraph 4.4.2 of MIL-M-38510?
24. Is PDA (percent defective allowable) established for Group *A, B,* or *C* Inspection per Method 5004 of MIL-STD-883?
25. How many times may a lot be submitted to burn-in testing? Are burn-in test condition and procedure (loading, frequency, and test circuit) specified? How often is what data to be recorded?

26. What is disposition of lot in case of burn-in or other lot-rejecting failure?
27. Must logistic spares or replacement parts be purchased from the same device class as original procurements?
28. Has a failure analysis procedure, including the minimum requirements for microsectioning, X-ray, and microphotography, been agreed upon?
29. Has Appendix *A*, entitled, "Product Assurance Program," of MIL-M-38510 been reviewed for application on specification as required?

30. Where automatic test equipment is utilized, such as Fairchild Models 4000, or 5000; or Texas Instruments 553 or HSM, are controlled standard test devices available to confirm proper operation at start of each lot? And to answer a question which arises about proper functioning of the test equipment: Are these correlation standard used to check or verify test equipment after calibration?

31. Are all the applicable items in this checklist clearly documented in the specification? Why not?
32. What (if any) tests are to be repeated by customer at receiving inspection/test?
33. Is supplier required to have an in-house reliability program per MIL-STD-883 (Method 5006) or equivalent?

Environmental Test Requirements

1. Which of the various test conditions in the following MIL-STD-883 test methods are applicable?

a.	Method 1001	Altitude	Method *A,B,C,* or *D*
b.	Method 1002	Immersion	Method *A,B,C,* or *D*
c.	Method 1004	Moisture Resistance	
d.	Method 1005	Steady State Life	
e.	Method 1006	Intermittent Life	
f.	Method 1007	Agree Life	Method *A,B,C,* or *D*
g.	Method 1008	High Temperature Storage	Method *A,B,C,* or *D*
h.	Method 1009	Salt Atmosphere (Corrosion)	
i.	Method 1010	Temperature Cycling	Method *A,B,C,* or *D*
j.	Method 1011	Thermal Shock	Method *A,B,C,* or *D*
k.	Method 1013	Dew Point	
l.	Method 1015	Burn-in	
m.	Method 2001	Constant Acceleration	Method *A,B,C,D, E, F,G,H,* or *J*
n.	Method 2002	Mechanical Shock	Method *A,B,C,D, E,F,* or *G*
o.	Method 2005	Vibration Fatigue	Method *A,B,C,* or *D*
p.	Method 2006	Vibration Noise	
q.	Method 2007	Vibration, Variable Frequency	Method *A,B,* or *C*

2. Is a continuity check made to assure test probes are still touching devices in *cold* test?
3. If CO2 is used to lower temperatures, is it admitted slowly into chamber to reduce "snow" which can adversely affect test connections? Is only solid form CO2 used in cold testing MOS ICs in order to prevent damaging static charges?
4. See also:
   - *a.* The series of "Environmental Test Checklist," Vol. VI.
   - *b.* "Test Procedure Checklist," Vol. VI.
   - *c.* "Troubleshooting and Problem Solving Checklist," 21.
   - *d.* "Failure analysis and Corrective Action Checklists," Vol. V.
   - *e.* All twenty-two "Product Assurance Checklists," Vol. V.

REFERENCES

"Microsystem Electronics Bulletin, No. 2"

"AR-5, Microelectronic Devices used in Avionics Equipment," Navy Department.

44. Microelectronic Process/Quality Checklist

W. B. Rossnagel*

This checklist was originally prepared in the Fall of 1967, when there were no formal government published specifications or standards to aid the microelectronic industry. It was written and distributed immediately to more than two dozen suppliers and more than one hundred persons actively engaged in the critical decisionmaking on process/quality problems without the benefit of accepted standards to measure against.

Six months after this checklist was originally released, the Rome Air Development Center released MIL-STD-883 which provided the needed standards for the microelectronics and integrated circuit industry. However, throughout 1968, a lesson in history repeated itself. That lesson was that despite a good standard there is still a need for very detailed checklists with which the engineers and process specialists can review the designs and specifications of the end product.

As a result of the continued requests for reprints of this checklist on the critical LIDS and Channels, which were state-of-the-art devices on the Minute Man III program, the author and his team revised and updated the document in the Spring of 1969. It was presented as a paper at the 8th Reliability and Maintainability Conference in Denver in July of that year.

To date, more than 750 copies of this checklist have been given to industry. It represents the ultimate in the detailed instructional and review aspects of a checklist written for a state-of-the-art process.

While many aspects of microelectronics were not covered in this checklist, it stands as a vivid example of how the checklist technique can be utilized to promptly identify, resolve, and control critical and expensive processes.

---

*The author wishes to acknowledge assistance by Ed Tighe, Dick Fellows, Ralph Fluharty, and Bill Schulze.

Abstract

This paper provides justification and recommendations for the process/quality controls needed in the manufacture, inspection, and test of today's military and space microelectronics and integrated circuits. It presents more than 250 detail points in checklist form for review and application as required to meet the requirements of these product lines.

The microelectronic processes are aptly illustrated in flowcharts. Matching checklists are provided for such operations as conductor screening, resistor abrading, and discrete part attachment. Several examples of visual aid accept/reject criteria are included to justify this essential function.

By reviewing and excerpting from this paper, engineering, manufacturing, and product assurance personnel have at their fingertips a comprehensive document which highlights the key areas of concern in microelectronic and IC manufacture.

No products ever manufactured in volume have required more stringent process and quality control than today's thick film microelectronic devices and integrated circuits (I.C.). When manufactured for military and space customers, these devices must pass extensive acceptance tests, sample environmental tests, and "burn-in" life tests, as well as maintaining log traceability controls and records.

The nature of building these devices, whether the manufacture is initiated by growing the silicon crystal or purchasing the wafer or die, causes an abnormally high percentage of cost to be expended in in-process assembly, inspection, and test. The normal ratio of having purchased parts and services represent 1/5 to 3/5 of an electronic product's manufacturing cost does not apply to these solid state devices. Most suppliers of microelectronics and integrated circuits report direct procurement costs of only five percent of the total product cost. This means that nominally ninety-five percent of all the "build costs" are expended within the plant in the form of many painstakingly critical operations. Some of these are preparation of photographic masters, compression bonding gold leads to die, and silk screening of several materials and epoxy applications. Recycling devices through manufacturing, inspection, and test sequences, holding lots for completion of sample nondestructive and destructive testing, designing, building, and maintaining expensive manufacturing and test facilities are also major contributors to the in-plant "build costs."

In microelectronic and I.C. devices, the insulating gaps between silk screened (thick film) conductors is often as close as .003" or .005". Most of the work is performed under magnifications from 10X up to 100X or higher in the case of photographic masters. Almost all of this work needs to be performed in clean rooms of class 100,000 (or better) per FED-STD-209.

A recent analysis of six suppliers providing the latest state-of-the-art LIDs (Leadless Inverted Devices), the very tiny devices shown in Fig. 3. showed the acceptable lot yield averaged only about twenty-five percent. On some lots, it was as low as ten percent, and scraping of entire lots, due to defective piece parts or out of control processes, was occurring frequently. This relatively low yield alone is cause for better control of the processes which cause rejections—and may affect reliability.

The process/quality recommendations learned from producing these pinhead size devices, which were not even conceived until late in 1968, is contained in the Section "LID and Channel Devices" of this working checklist. The first nine sections of this checklist are applicable to the buildup of all substrate microelectronic products.

One message is clear to those persons working in the microelectronic industry today, and it is particularly evident to the engineers and managers of General Electric who daily work with more than a dozen suppliers of completed substrates and modules—many of whom are producing identical devices. That message is this:

> Complete detailed documented process control procedures, plus a strong consistent and vigilant quality control organization provide the only way to produce high quality and high reliability microelectronic and I.C. devices.

The U.S. Air Force "Microelectronics Design Handbook" (Series 1-0) and "MIL-STD-883, Test Methods and Procedures for Microelectronics," are the initial publications providing the needed standards for this new industry (see references at end of checklist). However, these can only provide part of the criteria that is needed on the production floor. It is essential that additional definition, such as provided in this checklist, be written in manufacturing and QC procedures, accept/reject criteria, etc. General Electric has found it necessary to provide very detailed mil-spec drawing/specifications

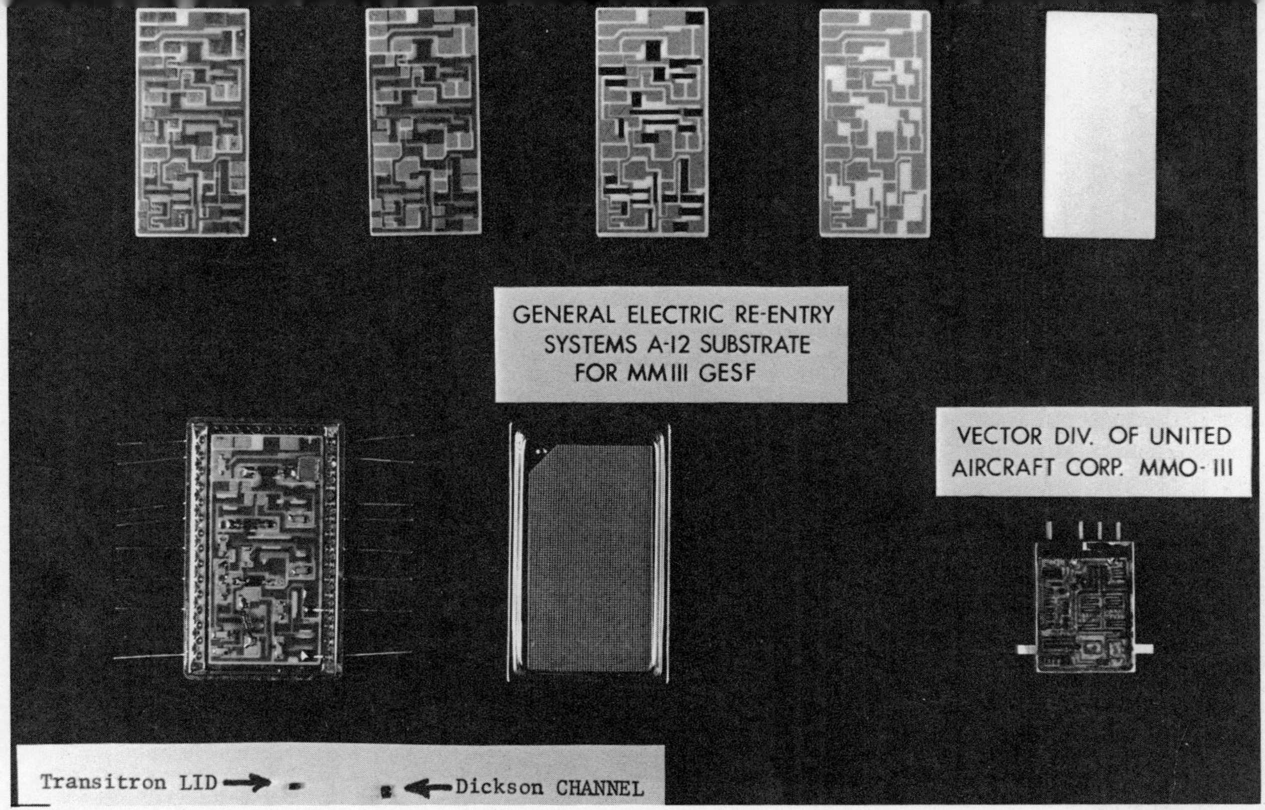

Figure 3. Microlectronic Devices Built on Substrates in Contract to an Integrated Circuit Device and Individual LID (Leadless Inverted Device) and Channel

as well as providing colored magnified accept/reject criteria, detailed checklists and working flow charts to aid its suppliers in its own in-plant operations.

Basic statistical limits and graphical records to indicate trends are indispensable tools to control the product to meet the quality/process specifications.

This paper is an example of a detailed checklist which can easily be incorporated as applicable, in the step-by-step procedures, planning, and visual aids for thick film manufacturing and quality control. More than 250 copies of the earlier version of this checklist, published by General Electric under "Technical Information Series," No. 68SD326, have been provided to government and industry. It also highlights to microelectronic design engineers the *realistic hardware-oriented engineering* definition that is necessary for the production of these devices.

It should be understood that this checklist cannot be all inclusive, nor can it be expected that all checklist items for this Hi-Rel military program will apply to every microelectronic supplier. In particular, commercial suppliers may build hardware to less stringent requirements. All these points are worthy of consideration by the suppliers, and many are applicable as the criteria for customer surveillance inspection checks and for determining receiving inspection characteristics.

However, the purpose of this checklist is to alert suppliers and customers to the many processes and quality controls which require consideration in the design, development, manufacture, inspection, and test of these microelectronic devices.

Alumina Substrate Elements

1. Are all substrates of standard sizes and thickness? Are any length or width tolerances tighter than ± .005"? ± .003"? Why?
2. Is alumina specified as unglazed? Is class and grade, per MIL-I-10, specified?
3. Is surface roughness specified as below 40 maximum per ASD standard B46.1?
4. Is maximum flatness (camber) less than .004"/in.?
5. Is a limit of hardness range established?

6. What size porosity is allowed under 10x magnification using oblique lighting? What quantity of what size checks, chips, scratches, and other imperfections is allowed? Is each defined?

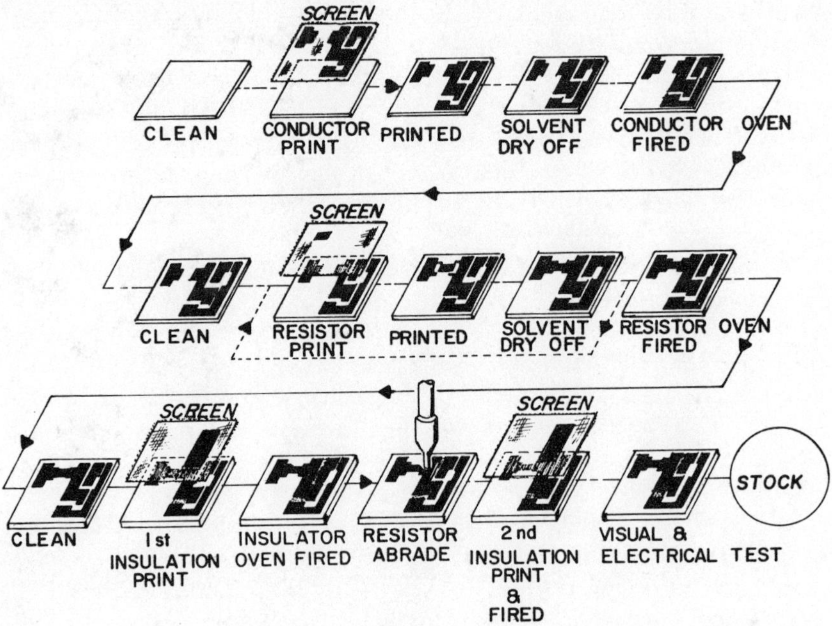

Figure 4. Thick-Film Substrate Fabrication Flow Diagram

7. Will cracks or chips more than .005" from edge be caused for rejection? If "shine through" technique is utilized, what is accept/reject criteria? Under what magnification?
8. Are all dimensional requirements met? Is thickness tolerance ± .002" or more?
9. Are squareness requirements met?
10. What allowable radius on edges is specified?

11. Is a binocular microscope used for receiving inspection? Is lighting on specimen 100 to 175 ft. candles? Is proper equipment being used for inspection of each attribute?
12. Are all substrates handled only with finger cots or bone tweezers or vacuum tweezers?
13. Does paperwork with substrates include as a minimum:
    - *a.* manufacturing name or symbol
    - *b.* part number
    - *c.* part name
    - *d.* lot number
    - *e.* date of manufacture
    - *f.* quantity
    - *g.* indication of acceptance?

14. When received and repacked for transfer to stockroom, are the individual substrates always kept in a closed container, and are they protected from each other?
15. Is the entire lot identified with "lot number"? Is individual substrate identification also necessary at this stage? Does at least one piece in any sample submitted for inspection or test represent the worst case in the lot?

16. Is each substrate and its container cleaned per details in written procedures?
17. Are requested samples submitted for inspection and approval before the entire lot is committed to screening? After screening?
18. Is a written requirement placed for detailed manufacturing and inspection planning at vendor? Can the procurement drawing or spec contain this detail?
19. Is an inspection record card initiated in receiving inspection to record acceptance, and reject number for each inspection parameter on each lot?
20. Does the inspection record card indicate what parameters are checked in the sample? On each substrate in the lot?

21. Is detailed planning provided for the receiving inspector?
22. How are substrates or carriers, which are plated by vendor, inspected upon receipt? How much may plating thickness vary over surfaces? Are sample microsections made to verify plating and thickness?

Conductor Screening

1. Is a dcsign review held on substrate layout? Is a breadboard model of circuit, using microelectronic parts, evaluated?
2. Are all notes on the drawing and process procedure reviewed and followed?
3. Is each substrate handled only by bone tweezers or finger cots throughout the fabrication process?
4. Has inspector recently reviewed manual before inspection prime hardware? Are visual aid accept/reject color photos in use?
5. Is the proper conductor screen used as per the latest revision of the bill of material and the drawing? Is double overlay screening allowed? Is rework screening allowed?
6. Was the conductor screen previously accepted? Does stamp show this?
7. Is a tradeoff study of quantity vs. thickness for film screen vs. emulsion screen made?
8. Is there a minimum of .003" between edge of all conductors and the edge of the substrate?
9. Is the registration of the conductor pattern to the substrate within .003"? Is there a minimum of .005" between conductors?
10. Is it understood that general visual inspection will be at 10x?
11. Is it understood that dimensional inspection of less than .020" will be at 30x? Is production using higher magnification than inspection?
12. Is it agreed that "holes" within a conductor will not exceed 10 per cent of a conductor width under 30x? Or within 50 per cent as in MIL-STD-883?
13. What is the agreement on the allowable waviness of the conductor edge?
14. Does any leeching of gold occur from out of the conductor, causing porosity? If yes, does the procedure provide sufficiently fine grain gold in the platinum gold paste?
15. Is the correct conductor paste used as per the B/M or the drawing?

16. Is the paste within shelf life period as per date on paste jar?
17. Are all cleaning solutions actually clean as per written requirements?
18. Is each substrate cleaned per details in written procedures?
19. Is the conductor paste mixed as per planning? Does a stamp show this? Is it kept in slowly rotating bottle rack?
20. Is each operation performed without undue lapse of time?

21. Are requested samples submitted for inspection and approval *before* the entire lot was committed to screening? After screening?
22. Is oven dry temperature-time profile maintained during lot drying time as per written specification? Does recorder show this?
23. Is furnace temperature stability checked as per written schedule each day?
24. Is lot number identification maintained? Is individual "piece part" identification also necessary at this stage? Does at least one piece in any sample submitted for inspection or test represent the worst case in the lot?
25. Were the required inspections made after the entire lot was fired? Cleaned? Identified? What was yield on opens? On shorts?

26. Are the proper number of substrates selected for sample solderability and adhesion tests?
27. Do the units selected for solderability and adhesion test perform satisfactorily? Was the lot accepted? Who reviewed data?
28. Have all rejects been separated from acceptable material and recorded on the shrinkage sheet? Are reworked units dot marked?
29. Are all substrates in proper handling containers? Is each container properly identified (including quantity) as per written procedures?
30. Is each operation, as required by the planning, and as shown on flow chart, performed on the entire lot?

Resistor Screening

1. Are all notes on the drawing and process procedure reviewed and followed?

TYPICAL CONDUCTOR PATTERN

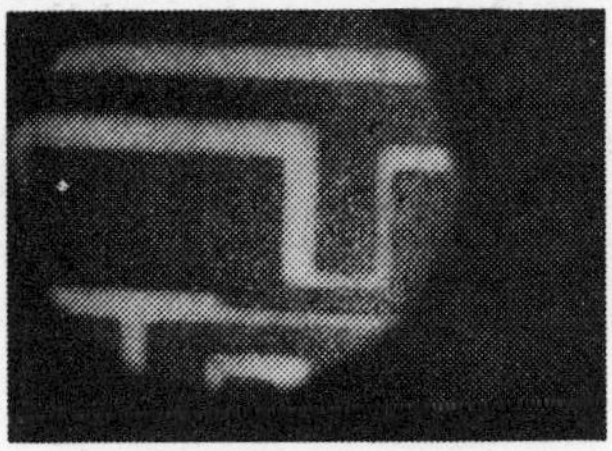

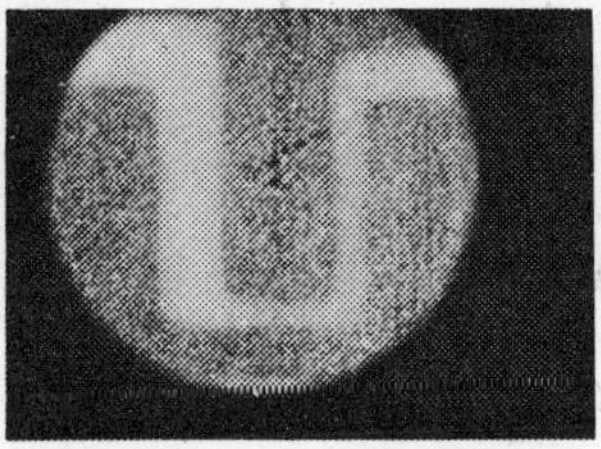

10X 20X

Above photo represents the inspection of a typical conductor pattern using a microscope at 10X & 20X for obvious defects.

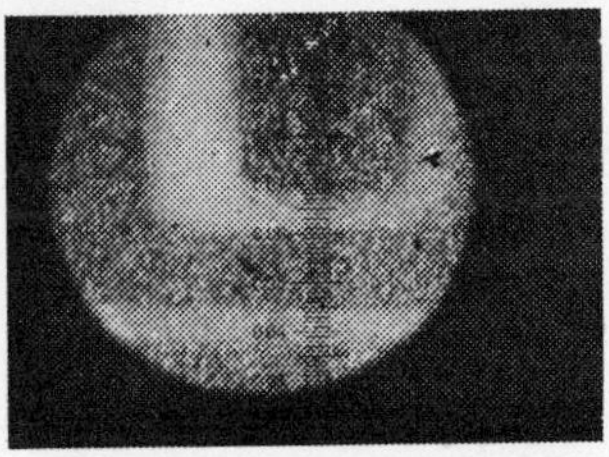

30X

Above photo represents inspection of a typical conductor pattern usinq a micrpscope at 30X and a reticle coordinate as the measure instrument.

NOTE: Above photographs are for general size comparison only.

Figure 5. Conductor Photographic Inspection Criteria

2. Is each substrate handled only by bone tweezers or finger cots throughout the fabrication process?
3. Has inspector recently reviewed manual before inspecting prime hardware? Are visual accept/reject color photos available?
4. Is the proper resistor screen used as per the latest revision of the bill of material and the drawing?
5. Was the resistor screen previously accepted? Does an inspection stamp show this?

6. Is the correct resistor paste or blend of pastes used as per the B/M and the drawing?
7. Is the paste within shelf-life period as per date on paste jar?
8. Is the resistor paste mixed as per planning? Is it kept in a slowly rotating bottle rack?
9. Is each substrate cleaned properly as per written procedures?

10. Is there a minimum of .003" between edge of all resistors and the edge of the substrate?

11. Does the resistor element overlap each conductor by a minimum of .006"?
12. Is the registration of the resistor pattern to the substrate within ±.003"?
13. Is there a minimum of .005" between resistors? Between resistor "ladder rungs" if used? Is there a minimum of .005" to any other deposited area for abrading purposes?
14. Is it understood that general visual inspection will be at 10x?
15. Is it understood that dimensional inspection of less than .020" will be at 30x?

16. Is it agreed that holes within a resistor will not exceed 10 per cent of the resistor width in the unbraided areas?
17. Is each operation performed without undue lapse of time?
18. Are requested samples submitted for resistance reading *before* the entire lot is committed to furnace firing?
19. Is the entire lot inspected during the air dry cycle?
20. Is oven temperature-time profile maintained during lot drying time as per written specification? Does recorder show this?

21. Is furnace temperature stability checked as per written schedule each day?
22. Is lot number identification maintained? Is individual piece part identification necessary at this stage? Does at least one piece in any sample submitted for inspection or test represent the worst case in the lot?
23. Are the required inspections made after the entire lot is fired, cleaned, and identified?
24. Are all rejects separated from acceptable material and recorded on the shrinkage sheet? Are reworked units dot marked?
25. Are all substrates in proper clean handling containers? Is each container properly identified (including quantity) as per written procedures? Are all parts free of damage at this point?

26. Is each operation, as required by the planning and as shown on flow sheet, performed on the entire lot?
27. If "shine through" method of inspection is used on resistors that do not overlap, what is criteria for accept/reject? Is shading

uniform? Is there a minimum of .006" overlap?

28. Is it agreed any imbedded lint, hairs, dust, or other foreign particles visible under 10x are cause for rejection before firing?
29. Is resistor photopositive dimensionally checked to with ± 001"? Is it compared to conductor drawing for compatibility? Was it superimposed on the prepared resistor screen?
30. Is resistor screening work performed in an adequate clean room? Does clean room meet class 100,000 or better of FED-STD-209?

Insulation (First Glass) Coat

1. Are all notes on the drawing and process procedures followed?
2. Is each substrate handled only by bone tweezers or finger cots throughout the fabrication process?
3. Are operator and inspector certified for this inspection?
4. Is the proper insulation screen used as per the bill of material and drawing?
5. Were the insulation screen and photopositive previously accepted? Does an inspection stamp show this?
6. What (if any) cracking or distortion is allowable after first glass under 10x? 20x? 30x? With direct and/or oblique lighting?
7. To avoid a "ping-pong" situation, is manufacturing building units to a greater magnification than inspection?
8. What (if any) size through voids are allowed over resistor elements after first glass? Under what magnification?
9. Will glass separation along resistor length be allowed if at least .002" from resistor?
10. Does glass cover any part of LID or channel pads?
11. Does glass cover all resistors on all sides by at least .002"? Is miniscus location established in accept/reject criteria?
12. are circular stains or "raindrops" cause for rejection?
13. How is contamination inherent in glass distinguished from controllable contaminants from air, people, tools, etc.?
14. Are all cleaning solutions filtered clean as per written procedures? Through what micron filters?
15. Is each substrate cleaned properly as per written procedures?
16. Is the correct insulation paste used as per the B/M or the drawing? Was it within shelf life?
17. Is insulation coating on lower conductor elements in crossovers

free of voids? Does coating overlap lower conductors by .010" minimum?

18. Is the insulation paste mixed as per planning? Is it kept in slowly rotating bottle rack?
19. Is each operation performed on the entire lot without undue lapse of time?
20. Is the entire lot inspected during the air cycle? Are all solder pads free of insulation coat?

21. Is oven dry temperature-time profile maintained during lot drying time as per written specification? Does recorder show this?
22. Is furnace temperature stability checked as per written schedule that day?
23. Is lot number identification maintained? Is individual piece part identification necessary at this stage?
24. Are the required inspections made after the entire lot is fired? Cleaned? Identified?
25. Since first glass inspection is last point where all previous operations can be seen, is it done 100 per cent? Under what magnification?

26. Is each operation, as required by the planning and as shown on flow sheet, performed on the entire lot?
27. Are all rejects separated from acceptable material and recorded on the shrinkage sheet? Are reworked units dot marked?
28. Are all substrates in proper handling containers? Is each container properly identified as per written procedures?

Resistor Abrading (Trimming)

1. Are resistor samples complete and abrading references established before entire lot is abraded?
2. Is the resistor value within 50 per cent of nominal value before abrading?
3. Is any resistor abraded more than 50 per cent across width within boundaries of connected conductors? Is digital ohmmeter data recorded for each resistor?
4. Does any resistor have less than 50 per cent contact at the terminating conductors?
5. Are control limits established within the allowable tolerance range?

6. Is it clear there shall be no conductor damage during abrading?

Photographic represents insulation coating voids on conductor element 30X.

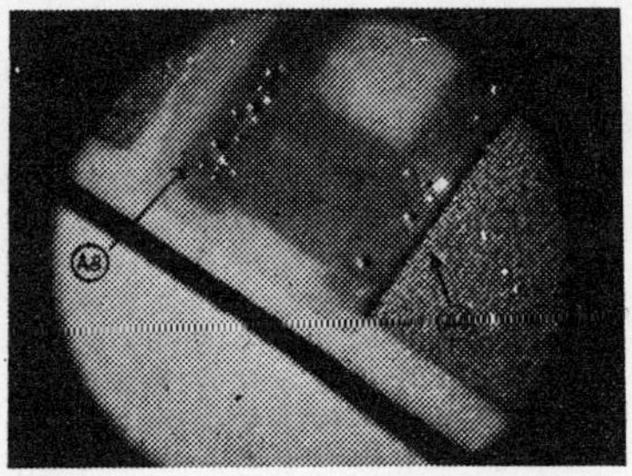

Reject Criteria

A8 - This photograph illustrates a condition where a large number of voids in the insulation coating are concentrated in one area over a conductor element. This condition was presented to the Process Control Engineer on the floor by the inspector. It was determined by the Process Control Engineer that this condition was non-acceptable. Voids in insulation coating covering conductor element are considered acceptable except where a large number of voids are concentrated in one area. The accept/reject criteria is determined by the Process Control Engineer who will determine if this condition is the result of a process out of control.

Accept Criteria

A4 - The glass insulating coat on the edge of solder pad area is considered even and undistorted.

Figure 6. Insulation Coat Photographic Inspection Guide

7. Is it agreed that no substrate shall be scored in any depth more than .003''?
8. Is abrading sand kept dry and unclogged by controlled drying elements? Is setting of pressure regulator checked daily?
9. Is nozzle diameter periodically checked? How often is it checked for pitting, cracking, or erosion?
10. Is feed control set as correct point? Is off-contact angle of nozzle correct?

11. Is a resistance measuring schematic and procedure established? Is it followed?
12. Is accuracy of digital ohmmeter adequate? Are probe tips rounded to prevent damage? Are pressure type tips used when possible?
13. Is effect of resistivity of solder pad areas included in low value readings?

14. Are all readings taken as close as possible to resistor? Is all equipment calibrated to MIL-C-45662A? Is a daily calibration check made with at least one standard resistor? Is this check included on data sheet?
15. Are all rejects separated from acceptable material and recorded on the shrinkage sheet? Are reworked units dot marked?
16. Are all substrates in proper clean handling containers? Is each container properly identified (including quantity) as per written procedures?
17. Is each operation, as required by the planning and as shown on flow chart, performed on the entire lot?

Note: See flow chart in Fig 7 for four phases of thick-film substrate operations:

1. substrate conductor screening
2. substrate resistor screening
3. substrate glass screening
4. solder tinning substrate conductors

Final Insulation (Second Glass) Coat

1. Are all notes on the drawing and process procedures followed?
2. Is each substrate handled only by bone tweezers or finger cots throughout the fabrication process?
3. Are operators and inspectors certified for this inspection?
4. Were the insulation screen and photopositive previously accepted?
5. Is any cracking or distortion visible under 10x? 20x? 30x? With direct and/or oblique lighting?
6. Is it agreed no through voids shall be over resistor elements? Any through voids are at least .002'' away from resistor? Are nonthrough voids exceeding .002'' over conductors acceptable?
7. Will glass separation along resistor length be allowed if at least .002'' from resistor?
8. Does glass cover all resistors on all sides by at least .002''?
9. Are circular stains or "raindrops" going to be rejected?
10. How is contamination inherent in glass distinguished from controllable contaminants from air, people, tools, etc.?
11. Are all conductor areas completely covered under 10x?

12. Are all cleaning solutions filtered clean as per written procedures?
13. Is each substrate cleaned properly as per written procedures?
14. Is the correct insulation paste used as per the B/M or the drawing? Was it within shelf life?
15. Is insulation coating on lower conductor elements in crossovers free of voids? Does coating overlap lower conductor by .010'' minimum?
16. Is a standard glass repair procedure established?
17. Does it describe criteria for homogeneous interface of new and old glass?
18. Are the proper number of substrates selected for temperature coefficient test? Is sample plan at least one out of every twenty-five?
19. Do the units selected for temperature coefficient test perform satisfactorily? Was the lot accepted?
20. Are all rejects separated from acceptable material and recorded on the shrinkage sheet? Are reworked units dot marked?
21. Are all substrates in proper clean handling containers? Is each container properly identified (including quantity) as per written procedures?
22. Has each operation, as required by the planning and as shown on flow sheet, been performed on the entire lot?

Substrate Soldering Fabrication and Inspection

1. Are the solvents clean? Are they properly applied?
2. Are the correct fluxes mixed properly? Are parts properly cleaned and at the correct temperature?
3. Is solder in pot clean? How is it "skimmed" clean without further contaminating it? Is pot corroded?
4. Is solder correct? Is solder "fresh"?
5. What is temperature gradient of solder pot? How often is controller calibrated? Is exact temperature recorded before each dip of each lot?
6. Is substrate cleaned within an hour before soldering?
7. What is temperature variation of preheating hot plate? Is it clean?
8. Is substrate heated at correct rate for correct time?
9. Is substrate cooled at correct rate for correct time?
10. Are speed and angle of entry and exit proper? Is cooling rate

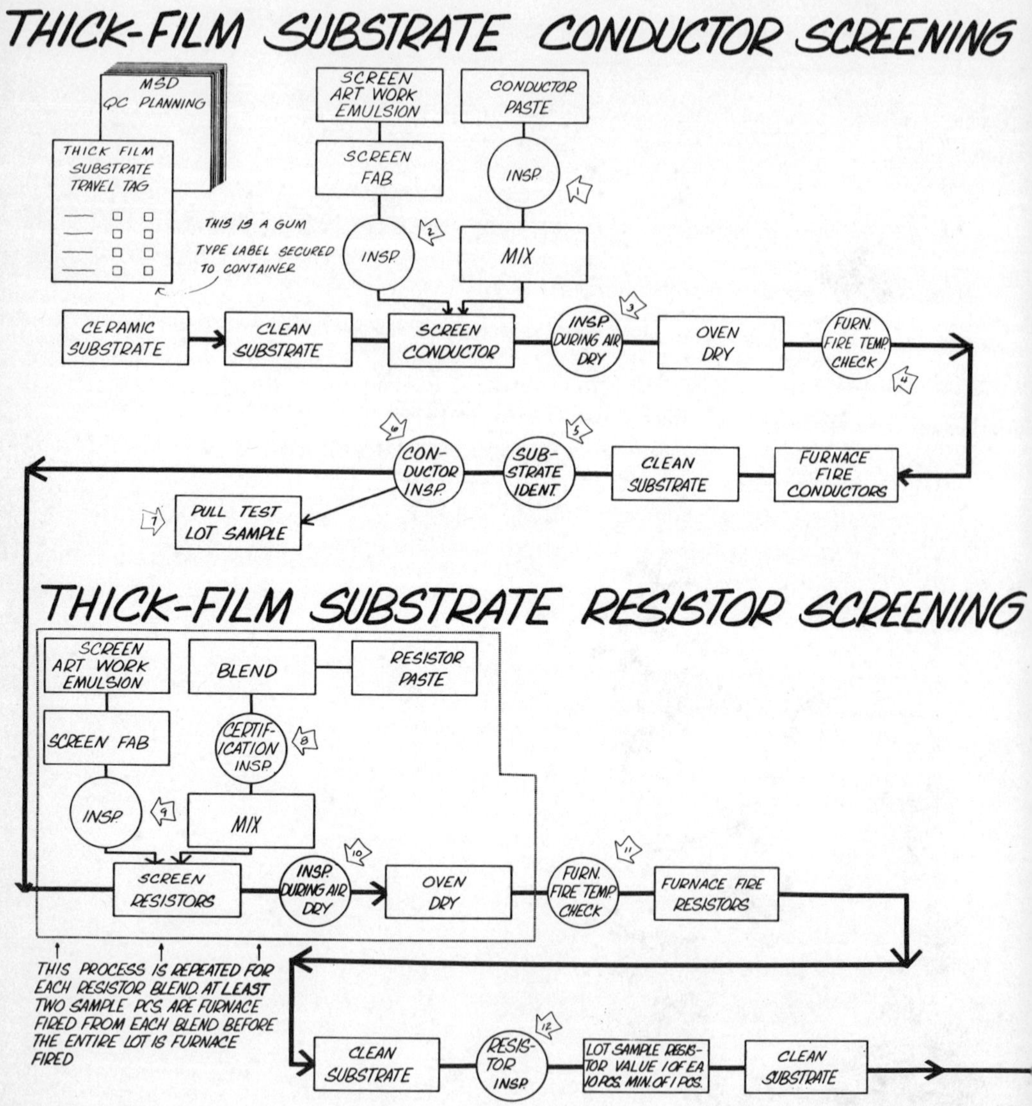
THICK-FILM SUBSTRATE CONDUCTOR SCREENING
MSD QC PLANNING
THICK FILM SUBSTRATE TRAVEL TAG
THIS IS A GUM TYPE LABEL SECURED TO CONTAINER
SCREEN ART WORK EMULSION
SCREEN FAB
INSP
CONDUCTOR PASTE
INSP
MIX
CERAMIC SUBSTRATE
CLEAN SUBSTRATE
SCREEN CONDUCTOR
INSP DURING AIR DRY
OVEN DRY
FURN. FIRE TEMP CHECK
FURNACE FIRE CONDUCTORS
CLEAN SUBSTRATE
SUB-STRATE IDENT.
CON-DUCTOR INSP.
PULL TEST LOT SAMPLE
THICK-FILM SUBSTRATE RESISTOR SCREENING
SCREEN ART WORK EMULSION
BLEND
RESISTOR PASTE
SCREEN FAB
CERTIF-ICATION INSP.
INSP.
MIX
SCREEN RESISTORS
INSP. DURING AIR DRY
OVEN DRY
FURN. FIRE TEMP CHECK
FURNACE FIRE RESISTORS
THIS PROCESS IS REPEATED FOR EACH RESISTOR BLEND. AT LEAST TWO SAMPLE PCS. ARE FURNACE FIRED FROM EACH BLEND BEFORE THE ENTIRE LOT IS FURNACE FIRED
CLEAN SUBSTRATE
RESIS-TOR INSP.
LOT SAMPLE RESIS-TOR VALUE 1 OF EA 10 PCS. MIN. OF 1 PCS.
CLEAN SUBSTRATE

Figure 7.

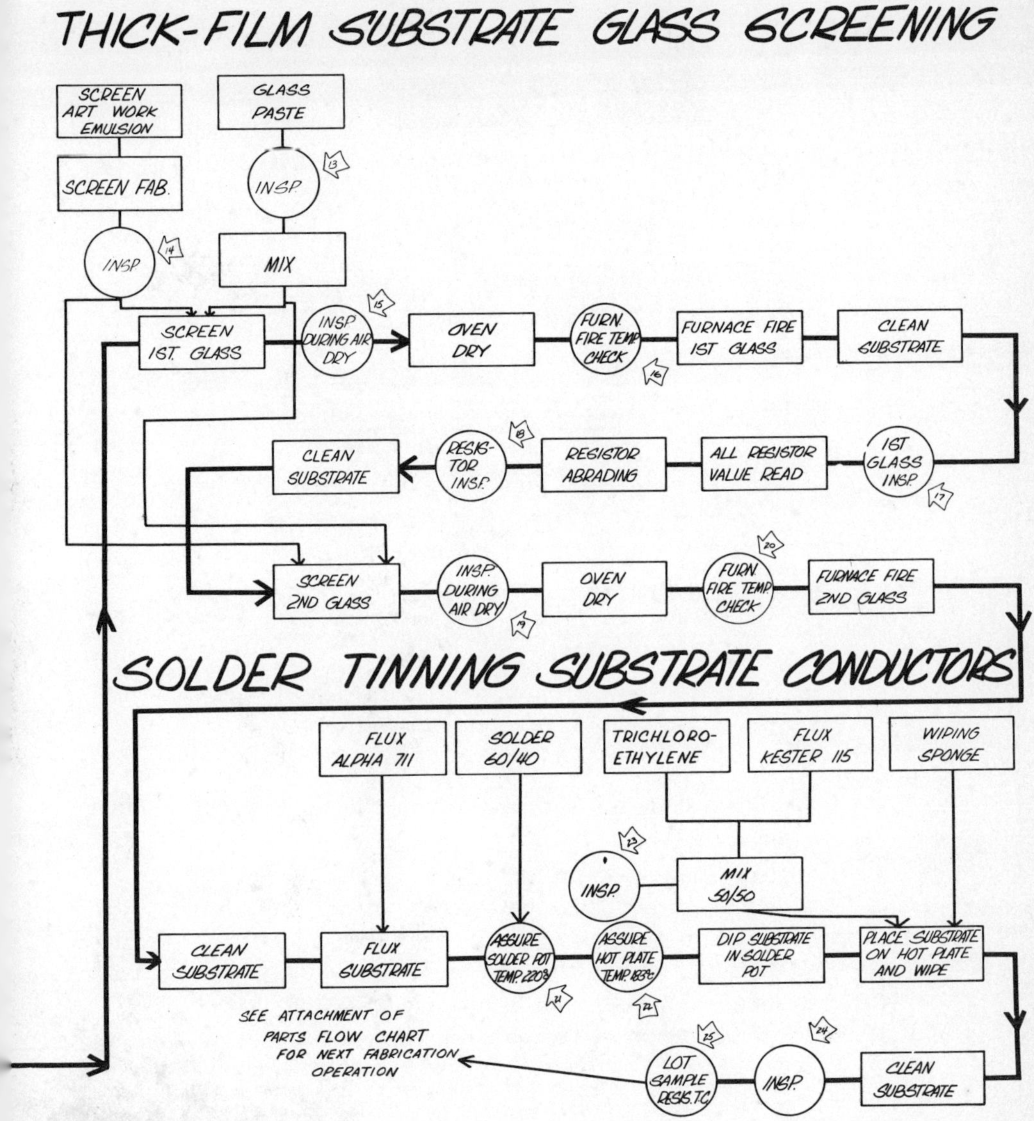
THICK-FILM SUBSTRATE GLASS SCREENING
SCREEN ART WORK EMULSION
GLASS PASTE
SCREEN FAB.
INSP.
INSP
MIX
SCREEN 1ST. GLASS
INSP. DURING AIR DRY
OVEN DRY
FURN. FIRE TEMP CHECK
FURNACE FIRE 1ST GLASS
CLEAN SUBSTRATE
1ST GLASS INSP.
ALL RESISTOR VALUE READ
RESISTOR ABRADING
RESISTOR INSP.
CLEAN SUBSTRATE
SCREEN 2ND GLASS
INSP. DURING AIR DRY
OVEN DRY
FURN. FIRE TEMP CHECK
FURNACE FIRE 2ND GLASS
SOLDER TINNING SUBSTRATE CONDUCTORS
FLUX ALPHA 711
SOLDER 60/40
TRICHLORO-ETHYLENE
FLUX KESTER 115
WIPING SPONGE
INSP.
MIX 50/50
CLEAN SUBSTRATE
FLUX SUBSTRATE
ASSURE SOLDER POT TEMP. 220°C
ASSURE HOT PLATE TEMP. 183°C
DIP SUBSTRATE IN SOLDER POT
PLACE SUBSTRATE ON HOT PLATE AND WIPE
SEE ATTACHMENT OF PARTS FLOW CHART FOR NEXT FABRICATION OPERATION
LOT SAMPLE RESIS. T.C.
INSP.
CLEAN SUBSTRATE

Figure 7.

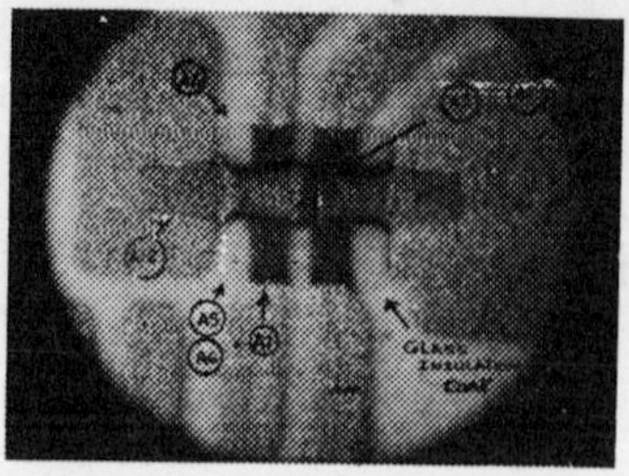

The inspector should realize the above photograph represents the end result of three separate operations:

1. Screening the lower conductor
2. Screening the insulated glass
3. Screening the upper conductor

The inspector shall use the "Shine-thru Method" for inspection of the lower conductor and the conventional microscope inspection at 30X for the cross-over (Upper Conductor). The glass insulation coat should be inspected for complete coverage of lower conductor prior to screening of the upper conductor.

Accept Criteria

A5 - Glass insulation coat should overlap the lower conductor by more than .010". In the above photo the actual distance was .020".

A6 - In the above photo the cross-over conductor was positioned with .025 inches glass coat on each side of cross-over conductor - Specification requires atleast .010 inches of glass on each side of the cross-over conductor.

A7 - Lower conductor in the above photo has width of .020 inches - Specification requires lowered conductor width to be at least .010 inches.

A12 - In the above photo the cross-over conductor element overlapped the conductor area by more than .006 inches. Specification requires the cross-over conductor element to overlap the conductor area by at least.006 inches.

Reject Criteria

A2 - A9 - The cross-over conductor at the points indicated was not uniform in shading and a close inspection revealed this area to be open.

Figure 8. Conductor Photographic Inspection Criteria, Conductor Crossover Net Work—20x

controlled by holding device over the surface of the solder pot for 4-5 seconds?

11. Is depth of immersion proper?
12. Is wiping operation performed as per written procedure?
13. Are only clean sponges used?
14. Is temperature variation of post dip hot plate correct? Is it clean?
15. Is cleaning of substrate performed as per written procedure?
16. Is inspection performed on entire lot?
17. Is there uniform solder flow on pad areas? Is wave soldering ramp angle correct?
18. Are pads completely covered?
19. Are solder surfaces clean and bright?
20. Is evidence of flux a cause for rejection and quarantine?
21. Is there a written procedure for the in-process removal of excess solder? Will paperwork show this?

22. Are all operators and inspectors fully qualified and trained in such techniques?
23. Are hardware samples of good and bad soldering available in addition to colored, visual pictures? Are these accept/reject criteria samples available to operators on the floor?
24. Will voids up to .002" (called pinholes) be accepted unless ten or more are concentrated in a .010 x .015" area? Will voids up to .006" be accepted unless two or more are concentrated within a .010" x .015" area?
25. Will voids more than .006" be cause for rejection?

26. Is a void defined as an area where the conductor material can be observed through the solder surface?
27. Is difference between voids and de-wetting agreed upon?
28. Is a 6-10 second immersion in solder pot used as a check for "leaching" of gold from pastes? Does it affect conductor porosity?
29. Is a solder immersion or leaching test established to insure no pad reduction occurs from tinning operation?

Discrete Part Attachment to Substrate

1. Are all notes on the drawing and process procedure reviewed and followed?
2. Is all handling of parts and substrate done with bone tweezers or finger cots?
3. Are all cleaning solutions checked? How? When? When discarded?
4. Is hot plate at proper temperature?
5. Is the correct flux ready for use?

6. Is solder pot at proper temperature?
7. Do all parts pulled per B/M show evidence of inspection? Are all discrete parts visually inspected under 20x? Do piece part vendors utilize accept/reject criteria to establish allowable cracks, delamination, lift off, etc.?
8. Are all parts pretinned? Cleaned?
9. Are applicable part leads performed as per drawing?
10. Is substrate properly prepared with no activated flux? Is flux quality proper?

11. Are parts properly placed on substrate? Are parts properly placed and secured?

12. Is substrate subjected to and removed from heat within required time limits?
13. After soldering, has the substrate been cleaned of all residual materials? Is a procedure established for this?
14. Is substrate inspected as per planning and visual aids?
15. Is the entire lot identified with "lot number"? Is individual substrate identification also necessary at this stage?

16. Does at least one piece in any sample submitted for inspection or test represent the worst case in the lot?
17. Are the proper number of substrates selected for solderability and adhesion test?
18. Are voids allowed in the tinned surface of a lead above the solder joint providing that surface is not part of the miniscus of the lead? What size voids? How many? Under what magnification?
19. Are interconnecting leads formed properly to prevent contact with covers? Will all leads be at least .015" below case lip?

Note: See Flowchart in Fig. 9 for final assembly substrate operations:
- *a.* attachment of parts
- *b.* placing substrate in case.

Substrate Mounting and Sealing

1. Is a careful study made of layout tolerance problems, alignment mismatch of substrate pads and glass bead terminals?
2. Are all substrates and cans being accepted by QC? Were glass headers inspected under 7X for crazing or cracks? Are lead through wires in center of glass headers? Were case and cover inspected for damage to plating?
3. Is case cover properly prepared as per planning?
4. What Mil-Spec do wires meet which connect header pins to substrate bonding pads?
5. Do all substrates, cases, and covers show evidence of acceptance by QC?

6. Are all mating surfaces really clean?
7. Is adhesive holding substrate properly prepared as per spec? To assure that adequate adhesive is applied, is it required that evidence of adhesive flow be visible at all edges of substrate?
8. Is adhesive cure cycle correct on substrate/case?

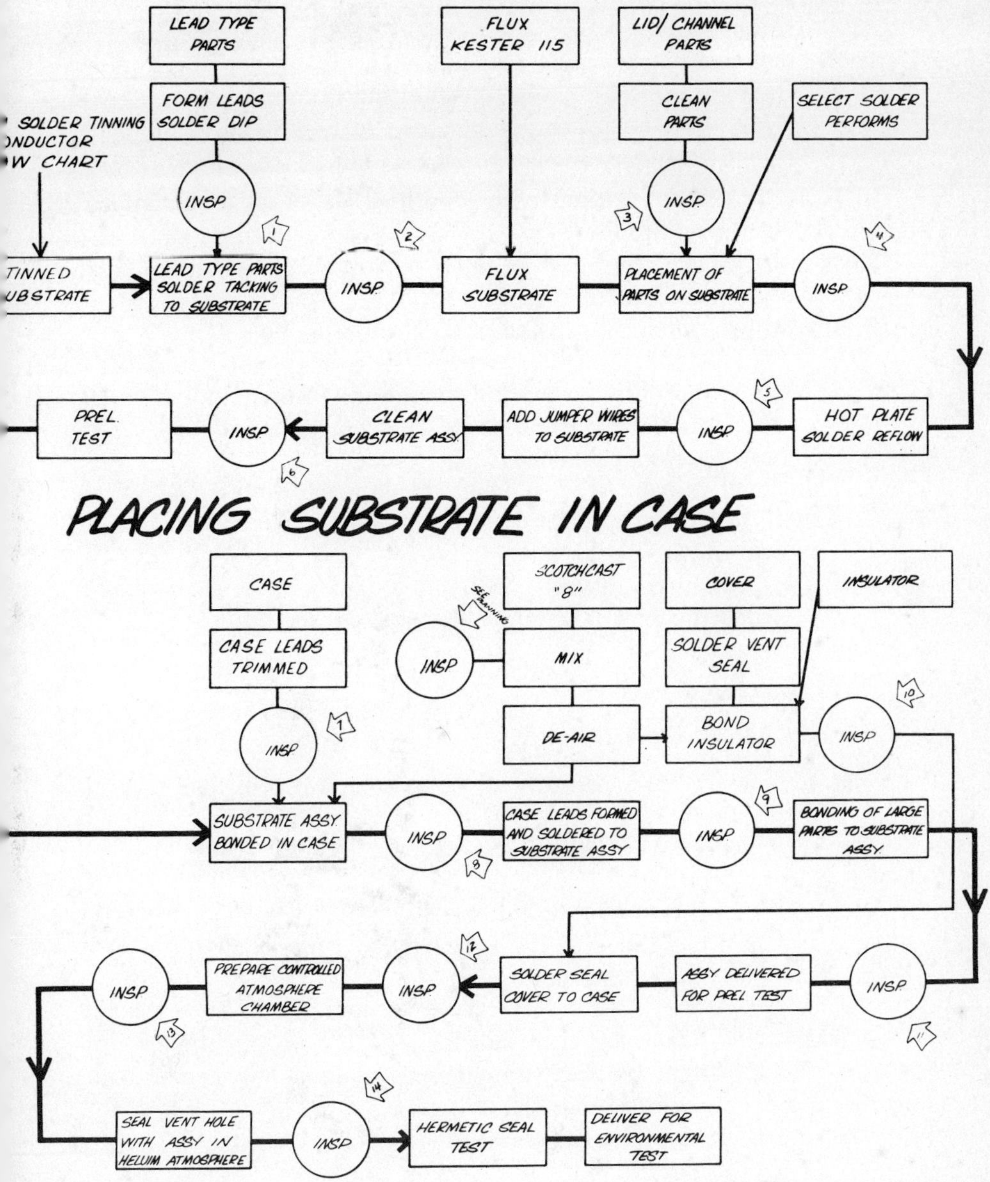

Figure 9. Attachment of Parts

9. Is unit final inspected for interface connections?
10. Is unit baked while being evacuated to remove all moisture?
11. Has unit been flushed and backfilled with dry N_2 - H_2 mixture?
12. Is solder, compression bond, or weld seal properly completed and inspected?
13. Is unit properly marked with identification, date of manufacture, part number, etc.
14. Is unit leakage checked? To what maximum allowable leakage rate? By what method?
15. If case is hermetically sealed, see checklist in Vol. III.
16. Is a procedure established for handling, testing, and inspecting hermetically sealed cases so as not to cause cracking of glass header lead beads?
17. How much crazing of glass beads is allowable under 7X?
18. Are all test fixtures carefully designed so as not to exert pressure on the glass beads?
19. Is an X-ray made to verify no "solder balls" have accumulated inside the module from solder sealing operation?

LID and Channel Devices

Note: *a.* LID (leadless inverted devices) contain up to four contact points
b. Channels contain only two contact points.

1. Encapsulant (Incoming Requirements)
   a. Does the drawing/specification contain all the necessary requirements?
   b. What inspections/tests are called out to be made on incoming encapsulants (epoxy type materials)?
   c. What are the storage requirements for encapsulant before mixing?
   d. Is shelf life defined?
   e. Is pot life defined?
   f. Is mixing process stringently controlled?
   g. After mixing, what controls are established on changing epoxy?
   h. Is epoxy allowed to be stored after mixing? Under what conditions?

2. Carrier (Substrate Requirements)
   - *a.* Are microsections or equivalent checks made to assure proper plating sequence and thickness? Is plating thickness checked by NDT?
   - *b.* Does data exist on distribution of plating thickness?
   - *c.* Are electrical leakage measurements performed? How many pico amps are allowed at what d.c. voltage?
   - *d.* Are applicable inspections of the first section of this checklist performed?
   - *e.* What sample plan is used? On what characteristics?
   - *f.* Is lot number identification maintained?
   - *g.* Are the cleanliness and handling procedures available where needed; are they being followed?
   - *h.* Is the isolation cut too deep?
3. Dice
   - *a.* Are dice processed in a class 100,000 or better clean room environment?
   - *b.* Are the cleanliness and handling procedures in place and being followed?
   - *c.* Is registration control adequate to prevent metallization misalignment?
   - *d.* Is precap visual inspection performed per MIL-STD-883?
   - *e.* Are the contact alignment and conductor scratch requirements established per MIL-STD-883?
   - *f.* What electrical tests are performed on the dice? On what sample plan?
   - *g.* Do the electrical tests performed on the wafer adequately qualify it for spec sheet or customer use?
   - *h.* Are the engineering and QC requirements specified on drawing adequate?
   - *i.* Is lot number identification maintained?
4. Die Bonding (Fastening Die to Carrier)
   - *a.* Are colored, magnified visual aids utilized that indicate the accept/reject criteria?
   - *b.* Is a shear test performed as per MIL-STD-883?
   - *c.* Is it a controlled test?
   - *d.* What do the in-process records indicate in terms of force applied to remove die? Will silicon die material fracture first?

- *e.* Are the in-process procedures and tests in place and being followed?
- *f.* Are the methods and criteria of MIL-STD-883 utilized?

5. Wire Bonding
   - *a.* Are colored magnified visual aids utilized for accept/reject criteria? What ratio range of wire bond to wire diameter is allowed?
   - *b.* What visual or mechanical criteria is provided operator on maintaining slack in the lead wire?
   - *c.* Do the temperature and pressure on the wire bond machine correlate with the written procedure for wire bonding?
   - *d.* Are pull tests being performed per MIL-STD-883? Daily sample, A.Q.L. or 100 per cent nondestruct?
   - *e.* What do the internal records indicate for lead bond pull strength, samples tested, area where break and/or breaks occur, etc.?
   - *f.* Is the critical station under adequate QC control?
   - *g.* Are visual inspection methods and criteria of MIL-STD-883 utilized?
   - *h.* Does the wire lead material meet MIL-STD-1267A?

6. Encapsulant Mix, Application, and Cure
   - *a.* Prior to using epoxy, are any checks made to ensure the use of only qualified epoxy?
   - *b.* Is epoxy carefully measured and mixed by lab personnel in a clean area?
   - *c.* Is epoxy dispensed in disposable type syringes? With manual or pneumatic pressure?
   - *d.* Is pot life of epoxy sufficient? Is a controlled amount placed on each device? If a large quantity (a few days supply) is stored, are the storage practices consistent with engineering procedures?
   - *e.* Does the operator jeopardize the die and/or die bonding quality when applying the encapsulant?
   - *f.* Does the cure cycle comply with the encapsulant manufacturer's recommended cure time and temperature?
   - *g.* What monitoring is performed during cure cycle?
   - *h.* Is the oven properly controlled?
   - *i.* What gauge is used to help measure epoxy height clearance?

j. What limits are placed on location of epoxy miniscus? Can a high miniscus be trimmed off?

k. What size holes, cracks, voids are allowed in epoxy under what magnification?

l. Is any separation allowed between vertical walls and epoxy?

m. Is lead wire protruding through epoxy?

7. Soldering

a. Are the proper cleaner, solder, and flux being used?

b. Are the cleaning procedures adequate? What is the maximum allowable time between cleaning and soldering?

c. What is the written schedule for changing cleaning solvents?

d. If a solder pot is used, does it contain at least two pounds of solder?

e. Is the dross removed frequently enough to insure good soldering? Is it skimmed off with a noncontaminating tool? Is a dust cover over the solder pot when not in use? Is a record kept of how often the solder is changed?

f. Is the solder pot itself clean?

g. What is the actual temperature variation of the solder pot?

h. What is the average number of times a part is solder-dipped to achieve acceptable solder criteria? If a petri dish is used atop a solder plate, is its solder discarded each day?

i. If a wave soldering operation is used, are all the variables specified and controlled?

j. What is the time and temperature the part is subjected to, i.e. are multiple dips required to attain the desired criteria?

k. Are the parts dipped consistently in the middle of the pot? To same depth?

l. Does all soldering exhibit a clean, bright appearance?

m. Is the solder process clean and tidy? Is there a data to indicate the process and operations are capable of successful repeatability?

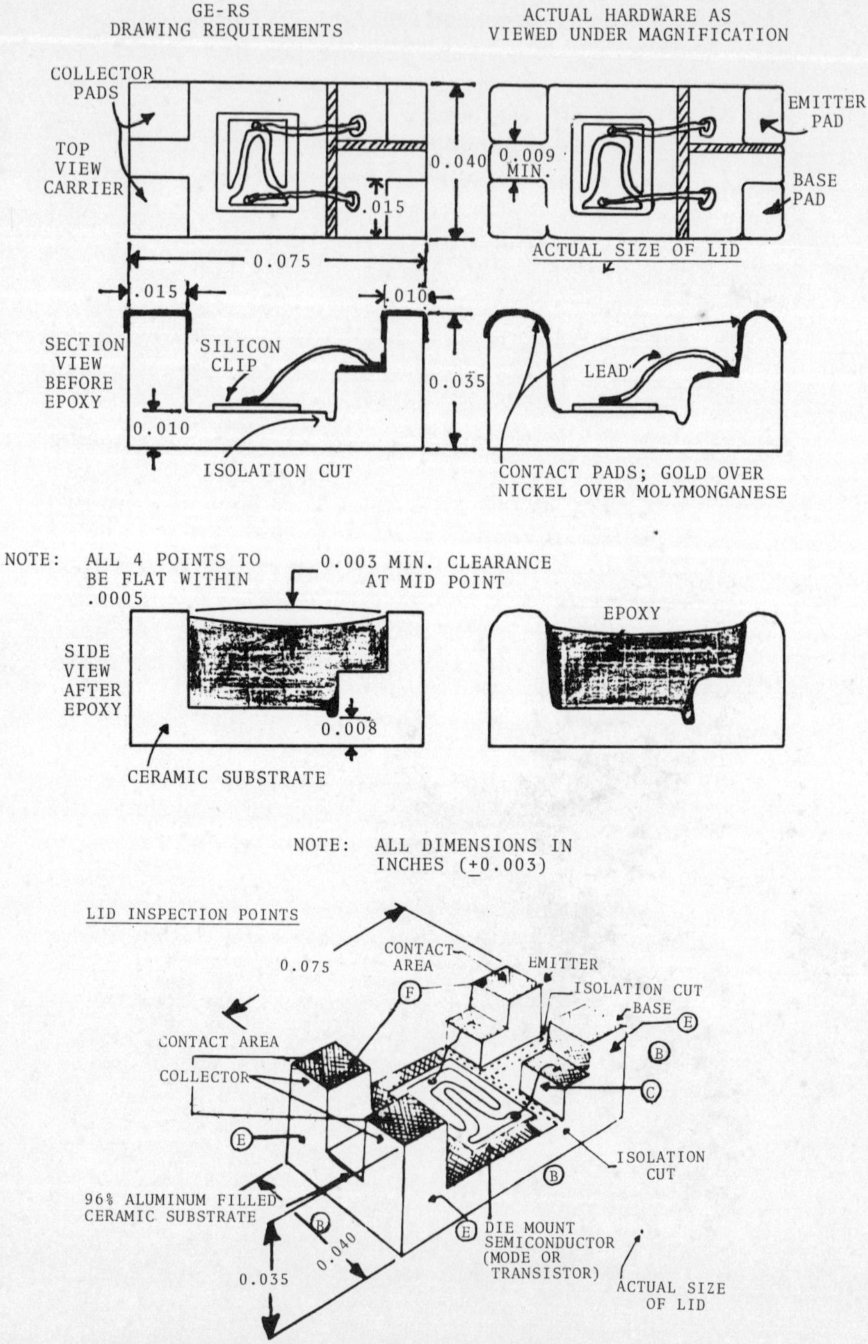

Figure 10. Microelectronics Lid (Leadless Inverted Device)

8. Electrical Test and Preparation for Shipment
   - *a.* Are the contact pins "floating" in the test fixtures in order to make adequate contact to assure the integrity of the connection?
   - *b.* Are only d.c. parameter measurements made?
   - *c.* Is there sufficient dwell time built into a specific parameter test so that the desired current level is achieved? How is this confirmed?
   - *d.* Does QC monitor or conduct acceptance tests?
   - *e.* Are tests conducted as per MIL-STD-883?

   - *f.* What is the vendor's experience on production test yields?
   - *g.* Is a failure investigation held if a considerable number of parts fail? What amount?
   - *h.* Are all devices kept in the proper cleaned and sealed containers? Is each package stamped off by QC?
   - *i.* Is a detailed certificate of compliance provided?
   - *j.* Are the parts correlated with test data sheets or tapes?

   - *k.* Does the device pass all the electrical tests? Does automatic test equipment monitor attribute limits or variables data?
   - *l.* Do all the parts meet the electrical parameters as required by drawing and specification? Does the travel card show this?
   - *m.* Is the part number (or code) marked properly on the specified flat surface of the carrier?
   - *n.* Does the shipping container include as a minimum: lot number, acceptance date code, part number, manufacturer's name, quantity, P.O. number?
   - *o.* Are the supplementary data and samples included in the shipment?

   - *p.* Are the temperature stabilization and/or thermal shock tests performed on a sample or 100 per cent basis?
   - *q.* Does the supplier run sample life tests from each lot which will indicate any long term degradation in performance?
   - *r.* Is it possible to trace a given lot back through the supplier's system?

REFERENCES

1. "AFSC DH 1-8, Microelectronics, General Design Series Handbook.
2. "FED-STD-151 Metals, Test Methods."
3. "MIL-F-14256 Flux, Soldering, Liquid (Rosin Base)."
4. "MIL-G-45204A, Gold Plating (Electrodeposited)."
5. "MIL-S-19491D, Semi-Conductor Devices, Packaging of."
6. "MIL-S-19500 Semiconductor Devices, General Specification for."
7. "MIL-S-45743, Soldering, Hi-Rel, Procedures for."
8. "MIL-S-6872A, Soldering Process, General Specification for."
9. "MIL-STD-883, Test Methods & Procedures for Microelectronics."
10. "MIL-STD-1276A, Weldable Leads for Electronic Component Parts."
11. "MIL-STD-1313 Microelectronic Terms & Definitions."
12. "N68-29525, Microelectronic Device Data Handbook (ARINC)."
13. "QQ-S-571 Solder, Tin Alloy, Lead Tin Alloy and Lead Alloy."